RELIABILITY AND MAINTAINABILITY IN PERSPECTIVE

Reliability and Maintainability in Perspective

Technical, management and
commercial aspects

David J Smith
BSc, C.Eng, FIEE, FIQA

First published 1981 by
THE MACMILLAN PRESS LTD
London and Basingstoke
Companies and representatives throughout the world

ISBN 0 333 31048 9 (hard cover)
ISBN 0 333 31049 7 (paper cover)

Typeset in 10/12pt IBM Press Roman by
STYLESET LIMITED
Salisbury · Wiltshire
and Printed in Hong Kong

Contents

6. Maintenance Philosophy and Down Time 50

6.1 Organisation of maintenance resources; 6.2 Maintenance procedures; 6.3 Tools and test equipment; 6.4 Personnel considerations; 6.5 Maintenance instructions; 6.6 Spares provisioning; 6.7 Logistics

7. Analysis of Failure Mode and Stress 59

7.1 Stress and failure; 7.2 Failure mode analysis; 7.3 Failure mechanisms; 7.4 Environmental stresses and failure rate; 7.5 Failure rate data

8. Design and Qualification Testing 69

8.1 Categories of testing; 8.2 Environmental testing; 8.3 Marginal testing; 8.4 High reliability testing; 8.5 Reliability growth; 8.6 Testing for packaging and transport; 8.7 Multiparameter testing; 8.8 Test houses

9. Quality Assurance and Automatic Test Equipment 76

9.1 Functions of QA; 9.2 Automatic test equipment

10. Maintenance Handbooks 89

10.1 The need for maintenance manuals; 10.2 A typical maintenance philosophy; 10.3 Information requirements for each group; 10.4 Types of manual; 10.5 Computer-aided fault finding; 10.6 The manual in perspective

11. Making Use of Field Feedback 95

11.1 Reasons for collecting field data; 11.2 Information to be recorded; 11.3 Difficulties involved; 11.4 Analysis and presentation of results; 11.5 Examples of failure report forms

Part III: Making Measurements and Predictions 103

12. Interpreting Data and Demonstrating Reliability 105

12.1 Inference and confidence levels; 12.2 The χ^2 test 12.3 Double-sided confidence limits; 12.4 Summarising the χ^2 test; 12.5 Reliability demonstration; 12.6 Sequential testing; 12.7 Setting up demonstration tests

Acknowledgements

I would particularly like to thank the following friends and colleagues for their help and encouragement with this book. Alex Babb, coauthor of *Maintainability Engineering* (Pitman 1973), for permission to quote freely from those pages and for his helpful suggestions concerning various aspects of the maintainability chapters.

Bernard Sharp, who has many years experience of reliability engineering with London Transport, for his very detailed study of the manuscript. His positive and helpful critique has played a significant part in finalising the style and layout.

Bruce Beach, of California USA, for his help with the software chapter and for permission to make use of some of his own material.

Len Nohre for a very thorough checking of the final manuscript and for his many helpful suggestions.

Brian Tilley, of Bristol Polytechnic, for comments on the chapter concerning product liability.

My wife, Margaret, for much help in preparing the manuscript.

I would also like to thank the Civil Aviation Authority and ITT Europe for permission to reproduce their failure report forms and the US Department of Defense for permission to quote from MIL Handbooks, in particular US MIL HDBK 217C.

Introduction

Reliability and Maintainability are already an essential part of design expertise demonstrated by more and more engineers being aware of such parameters as Availability, Down Time, Mean Time Between Failures and their economic importance in the finished product. This trend is largely due to the inclusion of Reliability and Maintainability requirements in the specifications called for by large national users of electronic and electro-mechanical equipment in particular the Ministry of Defence and the British Post Office. Since the middle 1960s these bodies, and many of the manufacturers of electrical and telecommunications equipment have invested in major reliability training schemes for their design, quality, production and management staff. The author has designed and run many such courses and management seminars for national and private organisations.

It cannot be overstated that satisfying such requirements is largely a matter of good engineering practice and the application of formal controls in design, manufacturing and service. The mathematical aspects of the subject, although important, serve only to refine measurement and do not themselves create a more reliable or more easily maintained product. Too often has the author had to discourage efforts to refine a reliability prediction or more precisely define a failure rate when an order of magnitude estimate would have sufficed. As with all aspects of engineering the ability to recognise the degree of accuracy required in a calculation and then to devise appropriate measurements is of the essence.

Reliability and maintainability are enhanced by the feedback of test and field-defect analysis, by the duplication of components and modules, by component selection and burn in and by the many management activities which it is the purpose of this book to outline. Very high costs of repair and the similarly high penalties which are incurred by expensive equipment being out of use, push reliability and ease of maintenance continually higher in the ranking of important design parameters. A single defect to a finished equipment, costs, more often than not, over £10 in diagnosis and replacement if it is detected in the factory, whilst the same fault in the field will likely cost upwards of £100 to rectify. An hour of down time of a communications link carrying several hundred telephone channels represents a lost revenue of at least £1000.

At the same time as escalating repair and down time costs increase the importance of these design parameters the task of achieving them is made more difficult by the complexity of the equipment. High function density components

(LSI chips) increase the possibility of obscure and elusive failures. The increasing use of computer control in the form of microprocessors, now used in many products from washing machines to petrol pumps, from telephones to motor cars, brings with it the possibility of equipment failure due to unforeseen behaviour of the software.

Ever decreasing development cycles, particularly for components, bring the complication that before a device is fully evaluated from field failure data a new generation has arrived to replace it. The more reliable the component the less failure data will present itself, thus aggravating the problem. In a rapidly changing technology the engineer has to accept a multi-disciplinary role in order to embrace reliability, maintainability, automatic test equipment, computer software in design, statistical methods, algorithmic maintenance instructions, legal and contract implications and many other topics. The following chapters aim to cover the very wide spectrum of activities and techniques involved in setting, measuring and achieving both reliability and maintainability objectives. The mathematical aspects are presented, not in erudite depth, but in as simple a way as possible consistent with imparting adequate knowledge and formulae for most practical purposes. Management and contractual aspects are covered and a chapter is devoted to a case study used for many years in training technical managers. Product Liability and Terotechnology are related topics of growing interest and are therefore included.

Since we are dealing with a number of engineering parameters, a practical, cost related, approach is essential. There will be a cost to achieve any parameter and a cost associated with each failure. Reliability and maintainability management aims to select and to achieve such levels of failure rate and repair time which keep the total of these costs to a minimum and only in this way can it justify its place in the spectrum of business activities.

Part I

Understanding Terms, Parameters and Costs

1 How Important are Reliability and Maintainability?

1.1 PAST AND PRESENT

Reference is often made in this type of literature to the spectacular reliability of many nineteenth-century engineering feats. Telford and Brunell indeed left a heritage of longstanding edifices such as the Menai and Clifton bridges. Fame is secured by their continued existence but little is remembered of the failures of their age. If, however, we concentrate on the success and seek to identify which characteristics of design or construction have given them a life span and freedom from failure far in excess of many twentieth-century products then two important considerations arise.

Firstly it is necessary to examine the nature of the comparison being made. The reliability of a structure or assembly will be influenced by its complexity. The fewer subassemblies and the fewer types of material and component involved then the greater is the likelihood of an inherently reliable product. The modern equipment and products which we so often condemn as unreliable are often comprised of thousands of piece parts involving many different materials all of which interact within various tolerances. Telford and Brunell's structures, on the other hand, are less complex comprising fewer types of material with relatively few well-proven modules.

Secondly we should consider the two most common methods of achieving reliability. They are:

DUPLICATION — The use of additional, redundant, parts whose individual failure does not cause the overall product to fail.
EXCESS STRENGTH — Deliberate design to withstand a higher stress than that which is required to be endured. Small increases in strength for a given applied stress result in substantial decreases in failure rate. This applies equally to mechanical and electrical designs.

Although effective, both are costly methods of achieving high reliability and long life. The next chapter will emphasise that reliability and maintainability are cost related and that the cost of any improvement in failure rate or repair time must be paid for by a reduction in operating or service costs or by increased revenue resulting from less down time.

The nineteenth-century engineers may not have been quite as prone to material cost constraints, or to the difficulties of equipment complexity, as are today's designers and this may account for much of the success of that age. No doubt many ventures did involve new materials and methods and were implemented under severe cost constraints. Perhaps they are the ones which have not survived to complete the comparison.

The purpose of the foregoing remarks is to point out that reliability is a 'built in' feature of any construction be it mechanical, structural or electrical and that it can be increased by design effort or by the addition of material. It is clear that the cost of such enhancement must be offset by at least the equivalent saving in maintenance in order to justify it. Maintainability is a related feature which determines repair times by a number of design features and maintenance methods and which must also be justified on a cost basis in the same way as reliability. Reliability and Maintainability together, at a given cost, dictate the proportion of time which the user will be able to use the equipment. The cost of ownership therefore will be that initial cost together with the cost of repair and the cost of lost usage resulting from the failures. It will be a recurring theme in this book that minimising this total is the basis of reliability and maintainability engineering.

1.2 REASONS FOR INTEREST

The substantial increase in importance attached to this subject over the last two decades is due partly to the dramatic increase in maintenance costs and partly to the difficulties inherent in complex equipment involving rapidly changing technologies. The following headings highlight the major reasons for this interest.

COMPLEXITY – Gives rise to intrinsic failures. These are failures not resulting from the clearly definable failure of a component part. They result from a combination of drift conditions or from unforeseen characteristics of software. They are hence more difficult to diagnose and less likely to be foreseen by the designer.
Results in a much larger number of possible failure modes. The number of ways in which an equipment failure may be caused is much greater in complex equipment thus making the task of prediction more prone to error.
MASS PRODUCTION – Requires a much higher degree of control over Material Procurement, Manufacture and Assembly, Engineering Changes and Concessions, etc. This type of production, with the division of labour involved, requires sophisticated systems of control and good Quality Assurance techniques in order to prevent manufacturing-related failures.
COST AND TOLERANCES – It is necessary to design to a production cost objective and this is often a severe restriction for commercial reasons. This leads to the calculation of tolerance and stress margins which will just meet the requirement. The probability of tolerance related failures in the field is thus increased.
Testing is now expensive and complex. Electronic test equipment can cost up

to £200 000 and test programming labour is costly. The temptation to prune
testing costs is often the cause of later failures.

MAINTENANCE — Field diagnosis and repair costs are much greater than those
  incurred in the factory. As a result reductions in failure rate and in repair time
  justify a reasonable investment.
  High complexity leads to the possibility of the maintenance activities
  themselves inducing failures as a result of faulty test equipment or human
  error.

1.3 ACTIVITIES INVOLVED

The achievement of reliability and maintainability results from activities in three
main areas:

DESIGN — Reduction in complexity. Use of standard proven methods.
  Duplication of modules to increase fault tolerance.
  Derating. This is the practice of using components of a higher stress rating
  than the minimum requirement.
  Prototype Testing sometimes called Qualification Testing.
  Subsequent feedback of all failure information into the design.
MANUFACTURE — Control of Materials, Methods, Changes, etc.
  Control of work standards (e.g. soldering) since these have long-term
  reliability effects.
FIELD SERVICE — Adequate Operating and Maintenance Instructions.
  Use of Preventive Maintenance including the elimination of Dormant Faults.
  Feedback of accurate failure information to design and manufacture (see
  chapter 11).
  Replacement Strategy.

The achievement of reliability and maintainability requirements involves, as can
be seen from the above list, a wide spectrum of management and engineering
activities. It should already be clear that they cannot be added after design and
manufacture by enhanced inspection and test but must be inherent in the design.
These parameters are part of the specification defining a product and can no
more be added later than can power consumption, weight, signal to noise ratio
or any other feature. In the event that this becomes necessary the cost is usually
prohibitive. No amount of MTBF (Mean Time Between Failures) calculations or
speculation, nor the use of more favourable figures, will ever enhance reliability.
The quest for more detailed failure rate information and its application to
reliability prediction is known as 'The Numbers Game'. It has its place in
reliability and maintainability engineering but is no more or less than a tool to
be used in a wider range of activities. Actual improvements are only achieved by
the application of the activities mentioned above and covered in more detail in
chapter 4.

It has been stressed that we are discussing design parameters which, therefore, should be chosen in advance. Design and Development should then follow and the equipment then evaluated against the original objectives. It follows that the inherent reliability and maintainability are determined, in the first instance, by the 'drawings'. This results from the inherent reliability of the chosen components, their quantity, method of interconnection and from other aspects of the equipment practice. We shall refer to this as the design reliability and it is this value which characterises the equipment although it is never achieved in practice. The activities of manufacture, operation and service then give rise to failure possibilities and the achieved reliability and maintainability are always less than the theoretical design level. Reliability activities tend, therefore, to fall into two areas covering design and assurance. Figure 1.1 emphasises this philosophy in the form of a diagram showing design reliability with some of its associated activities and achieved reliability, at a lower level, with some of the assurance activities.

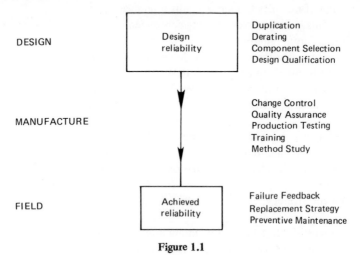

DESIGN — Design reliability — Duplication / Derating / Component Selection / Design Qualification

MANUFACTURE — Change Control / Quality Assurance / Production Testing / Training / Method Study

FIELD — Achieved reliability — Failure Feedback / Replacement Strategy / Preventive Maintenance

Figure 1.1

Reliability and Maintainability are interdependent for three reasons.

1. A system whose reliability is partly dependent on a degree of duplication (redundancy) will be more reliable if the repair time (maintainability) of failed redundant units is improved. Maintainability is therefore capable of contributing to reliability.
2. The design and assurance activities required to achieve these parameters are, in many cases, the same.
3. Both parameters contribute to the overall AVAILABILITY, that is to say available UP TIME, of the product.

AVAILABILITY now emerges as a useful concept. In many cases it more accurately defines the user requirement than either reliability or maintainability alone. It is achieved by a combination of the two and permits a trade off between them to achieve a given level of availability. An equipment failing once every 9 days and taking one day to repair has an availability of 90 per cent. So also does an equipment failing once every 18 days and taking 2 days to repair. In chapter 3 emphasis will be given to the choice of appropriate parameters when describing a requirement. Availability is often the most appropriate and useful parameter.

1.4 CONTRACTUAL PROBLEMS

The reasons for the upsurge in interest in this subject have now been stated and it is these which have resulted in reliability and maintainability requirements becoming commonplace in contracts. Specified MTBFs and repair times are to be found in many technical specifications and it will be no surprise to learn that these are frought with pitfalls arising from:

Ambiguity of definitions
Hidden statistical risks
Inadequate coverage

These pitfalls will be covered in detail in chapter 18. Requirements are called for in two ways:

BLACK BOX SPECIFICATION – Suitable for stating a reliability or
    maintainability requirement for a low, or medium reliability module required
    in reasonable quantities. Statistical demonstration is therefore appropriate.
TYPE APPROVAL METHOD – Applicable to complete systems with long
    development cycles and involving small quantities. Demonstration of reliability
    is, in such cases, not feasible and customer involvement, at all stages of design
    and manufacture, is the only means of assuring that it is achieved.

In practice a combination of these two approaches is used. The Type Approval method, together with customer controls and involvement in the various stages of over-design and therefore overstating the requirements will prove to be expensive. Objectives must be chosen on the optimum cost basis already discussed. These will only be chosen, and economically achieved, if the overall issues raised by reliability and maintainability are fully understood.

2 A Realistic Approach is Cost Conscious

2.1 COST OF QUALITY AND RELIABILITY

The practice of identifying quality costs is not new although it is only very large organisations that collect and analyse this highly significant proportion of their turnover. Attempts to set budget levels for the various elements of quality costs are even rarer as is the planning of activities for achieving them. This is unfortunate since the contribution of any activity to a business is measured ultimately in financial terms and the activities of quality, reliability and maintainability can claim no exception. If the costs of failure and repair were more fully reported and compared with the costs of achieving improvements then greater strides would be made in improving the position of this branch of engineering management. Greater recognition leads to the allocation of more resources. The pursuit of quality and reliability for their own sake is no justification for the investment of labour, plant and materials. Value Engineering, Work Study, Computer Planning, and other functions are quick to demonstrate that the savings generated by their activities more than offset the expenses involved.

A Quality Cost analysis entails extracting various items from the accounts and grouping them under the three headings:

PREVENTION COSTS — Costs of preventing failures.
APPRAISAL COSTS — Costs related to measurement.
FAILURE COSTS   — Costs incurred as a result of scrap, rework, failure, etc.

Each of these categories can be broken down into identifiable items and table 2.1 shows a typical breakdown of quality costs for a six-month period in a manufacturing organisation. The totals are expressed as a percentage of sales, this being the usual ratio used. It is known by those who collect and study these costs that they are usually under-recorded and that the failure costs indicated may be as little as three-quarters of the true value incurred. The ratios shown in table 2.1 are typical of a manufacturing and assembly operation involving light machining, assembly, wiring and functional test of electrical equipment. The items are as follows.

Table 2.1 Quality costs

1 Jan 1979–30 June 1979

(sales £2 000 000)

Prevention costs	£ , 000	% of Sales
Design review	0.5	
Quality and reliability training	2	
Vendor quality planning	2.1	
Audits	2.4	
Installation prevention activities	3.8	
Product qualification	3.5	
Quality engineering	3.8	
	18.1	0.91
Appraisal costs		
Test and inspection	45.3	
Maintenance and calibration	2	
Test equipment depreciation	10.1	
Line quality engineering	3.6	
Installation testing	5	
	66.0	3.3
Failure costs		
Design changes	18	
Vendor rejects	1.5	
Rework	20	
Scrap and material renovation	6.3	
Warranty	10.3	
Commissioning failures	5	
Fault finding in test	26	
	87.1	4.36
Total quality cost	171.2	8.57

Prevention costs   Design Review — Review of new designs prior to the release of drawings.
Quality and Reliability Training — Training of QA staff. Q and R Training of other staff.

Vendor Quality Planning — Evaluation of vendors' abilities to meet requirements.
Audits — Audits of systems, products, processes.
Installation Prevention Activities — Any of these activities applied to installations and the commissioning activity.
Product Qualification — Comprehensive testing of a product against all its specifications prior to the release of final drawings to production. Some argue that this is an appraisal cost. Since it is prior to the main manufacturing cycle the author prefers to include it in Prevention since it always attracts savings far in excess of the costs incurred.
Quality Engineering — Preparation of quality plans, workmanship standards, inspection procedures.

Appraisal costs   Test and Inspection — All line inspection and test activities excluding rework and waiting time. If the inspectors or test engineers are direct employees then the costs should be suitably loaded. It will be necessary to obtain, from the cost accountant, a suitable overhead rate which allows for the fact that the QA overheads are already reported elsewhere in the quality cost report.
Maintenance and Calibration — The cost of labour and subcontract charges for the calibration, overhaul, upkeep and repair of test and inspection equipment.
Test Equipment Depreciation — Include all test and measuring instruments.
Line Quality Engineering — That portion of quality engineering which is related to answering test and inspection queries.
Installation Testing — Test during installation and commissioning.

Failure costs   Design Changes — All costs associated with engineering changes due to defect feedback.
Vendor Rejects — Rework or disposal costs of defective purchased items where this is not recoverable from the vendor.
Rework — Loaded cost of rework in production and, if applicable, test.
Scrap and Material Renovation — Cost of scrap less any reclaim value. Cost of rework of any items not covered above.
Warranty — Warranty: labour and parts as applicable. Cost of inspection and investigations to be included.
Commissioning Failures — Rework and spares resulting from defects found and corrected during installation.
Fault Finding in Test — Where test personnel carry out diagnosis over and above simple module replacement then this should be separated out from test and included in this item. In the case of diagnosis being carried out by separate repair operators then that should be included.

A study of the above list shows that reliability and maintainability are directly related to these items.

The Department of Prices and Consumer Protection document entitled 'A

National Strategy for Quality' (1978) estimates that UK Industry turnover for 1976 was £105 thousand million. The total quality cost for a business is likely to fall between 4 and 15 per cent, the average being somewhere in the region of 8 per cent. Failure costs are usually approximately 50 per cent of the total — higher if insufficient is being spent on prevention. It is unlikely then that less than £4.2 thousand million was wasted in defects and failures. A 10 per cent improvement in failure costs would release approximately

£500 MILLION

into the economy. Prevention costs are likely to be approximately 1 per cent of the total and therefore £1 thousand million. A mere 5 per cent increase (£50M) might well achieve the £500 million saving in failure.

2.2 INTRODUCING A QUALITY COST SYSTEM

Convince top management — Initially a quality cost report similar to table 2.1 should be prepared. The accounting system may not be arranged for the automatic collection and grouping of the items but this can be carried out on a one off basis. The object of the exercise is to demonstrate the magnitude of quality costs and to show that prevention costs are small by comparison with the total.

Collect and Analyse Quality Costs — The data should be drawn from the existing accounting system and no major readaption should be made. In the case of change notes and scrapped items the effort required to analyse every one may be prohibitive. In this case the total may be estimated from a representative sample. It should be remembered, when analysing change notes, that some may involve a cost saving as well as an expenditure. It is the algebraic total which is required.

Quality Cost Improvements — The third stage is to set budget values for each of the quality cost headings. Cost improvement targets are then set to bring the larger items down to an acceptable value. This entails making plans to eliminate the major causes of failure. Those remedies which are likely to realise the greatest reduction in failure cost for the smallest outlay should be chosen first.

2.3 USER QUALITY COSTS

So far only manufacturers' quality costs have been discussed. The costs associated with acquiring, operating and maintaining an equipment are equally relevant to a study such as ours. The total costs incurred over the period of ownership of an equipment are often referred to as LIFE CYCLE COSTS. These can be separated into:

Acquisition Cost — Capital cost plus cost of installation transport, etc.

Ownership Cost — Cost of preventive and corrective maintenance and of
    modifications.
Operating Cost — Cost of materials and energy.
Administration Cost — Cost of data acquisition and recording and of
    documentation.

They will be influenced by:

Reliability — Determines frequency of repair.
    Fixes spares requirements.
    Determines loss of revenue (together with maintainability).
Maintainability — Affects training, test equipment, down time, manpower.
Safety Factors — Affects operating efficiency and maintainability.

Life cycle costs will clearly be reduced by enhanced reliability, maintainability
and safety but will be increased by the activities required to achieve them. Once
again the need to find an optimum set of parameters which minimises the total
cost is indicated. This concept is illustrated in figures 2.1 and 2.2. Each curve
represents cost against Availability (calculated from Reliability and Maintainability, chapter 3). Figure 2.1 shows the general relationship between availability
and cost. The manufacturer's pre-delivery costs, those of design, procurement
and manufacture, increase with availability. On the other hand, his after-delivery
costs, those of warranty, redesign, loss of reputation, decrease as availability
improves. The total cost is shown by a curve indicating some value of availability
at which minimum cost is incurred. Price will be related to this cost. Taking,

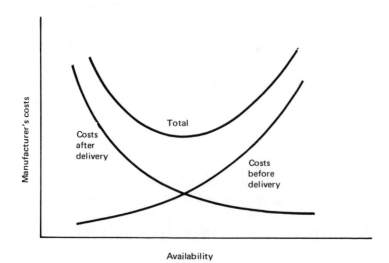

Figure 2.1   Availability and cost — manufacturer

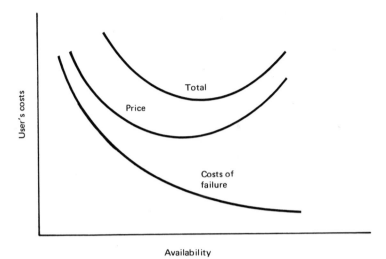

Figure 2.2   Availability and cost – user

then, the Price/Availability curve and plotting it again in figure 2.2 the user's costs involve the addition of another curve representing losses and expense, due to failure, borne by him. The result is a curve also showing an optimum availability which incurs minimum cost. Such diagrams serve only to illustrate the philosophy whereby cost is minimised as a result of seeking reliability and maintainability enhancements whose savings exceed the initial expenditure.

2.4 COST AND PERFORMANCE

In practice a number of factors influence the choice of an equipment as is shown in figure 2.3.

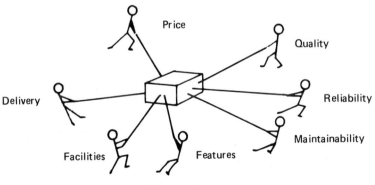

Figure 2.3

Their relative importance will depend on the circumstances. Delivery could be the overriding reason for choosing a particular manufacturer even if the other factors are inferior to the competition. On the other hand reliability and maintainability might emerge as the most important if loss of revenue or of life could result from failure. In the absence of special circumstances favouring one factor only the user will attempt to choose a product with the mix of factors best suited to his requirements. An enhancement of one feature may be possible at a greater cost or by the reduction of some other parameter. Greater reliability for less maintainability or less reliability in return for more facilities are examples of these trade-offs.

2.5 RELATIVE DEFECT COSTS

The cost of finding and rectifying defects increases dramatically as an item moves through the stages of manufacture and installation. Typical costs resulting from the same defect discovered at various stages are:

Component at Incoming Inspection, say	10p
Same Component in a Printed Board, say	£1.00
In the Finished Product under test, say	£10.00
In the Product in Field Use, say	£100.00

This is due to maintainability considerations such as Access and Diagnosis which will be discussed in chapter 5. The aim is to detect and rectify failures as early as possible in the manufacturing cycle. It is on this basis that capital for incoming component testing is usually justified.

2.6 THE COMPLEX EQUATION

The cost of providing R and M, the relationship between them, and the cost of maintenance, form a complex interaction about which it is difficult to generalise. Money spent on maintainability reduces repair time, which in turn enhances reliability in the presence of redundant units. Improved reliability reduces maintenance costs whereas money spent on preventive maintenance may enhance reliability. Preventive and corrective maintenance are related as shown by figure 2.4.

Increased preventive maintenance brings down repair costs and the sum of the two falls. There is, of course, a trade-off beyond which an increase in preventive maintenance no longer produces an equivalent fall in repair cost. Trade-offs are made between the many parameters involved according to requirements and circumstances. Clearly it is not possible to optimise every parameter nor to calculate costs from a magic equation which takes account of every one. The approach to the possibility of improving any parameter should always be to assess the cost of an increment in that parameter and compare that with the savings which will accrue if that increment is achieved.

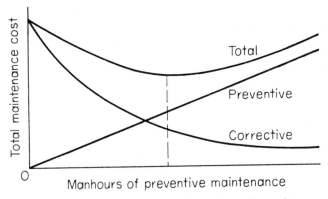

Figure 2.4   Trade-off between preventive and corrective maintenance

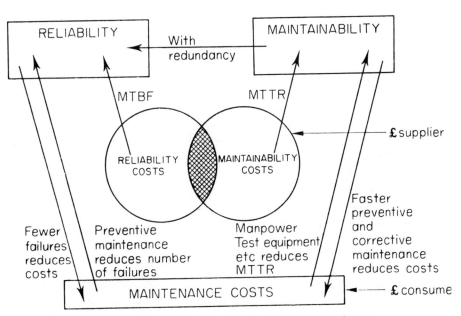

Figure 2.5   Interrelationship of system effectiveness and costs

3 Understanding Terms and Jargon

3.1 QUALITY, RELIABILITY AND MAINTAINABILITY

Before any discussion involving these terms can take place it is essential that the word FAILURE is fully defined and understood. Unless the failed state is defined it is impossible to explain the meaning of Quality or of Reliability. There is only one definition of failure and that is:

NON-CONFORMANCE TO SOME DEFINED PERFORMANCE CRITERION

Refinements of definitions which differentiate between terms such as Defect, Malfunction, Failure, Fault and Reject are important in contract clauses and in classification and analysis of data but should not be allowed to cloud the understanding of the main parameters. The different definitions of these terms merely include and exclude failures by type, cause, degree or use. Given any specific definition of failure there is no ambiguity in the definitions of quality and reliability. Since failure is defined as departure from specification then to define different types of failure implies the existence of different performance specifications. Table 3.1 gives an indication of the classification of failures.

The much used Bathtub Curve is an example of the practice of treating more than one failure type by a single classification. It seeks to describe the variation of Failure Rate of electrical components during their life. Figure 3.1 shows this generalised relationship as it is assumed to apply to electronic components. The failures exhibited in the first part of the curve, where failure rate is decreasing, are called early failures or infant mortality failures. The middle portion is referred to as the useful life and it is assumed that failures exhibit a constant failure rate, that is to say they occur at random. The latter part of the curve describes the wearout failures and it is assumed that failure rate increases as the wearout mechanisms accelerate.

Figure 3.2, on the other hand, shows the bathtub curve to be the sum of three separate overlapping failure distributions. Labelling sections of the curve as wearout, burn in and random can now be seen in a different light. The wearout region implies only that wearout failures predominate, namely that such a failure is more likely than the other types. The three distributions are as in table 3.2.

Table 3.1

By cause	Stress-related failure
	Misuse failure
	Inherent weakness failure
	Wearout failure
	Maintenance induced failure
By suddenness	Immediate failure
	Gradual degradation failure
By degree	Catastrophic failure
	Intermittent failure
	Partial failure
By definition	Applicable to the specification
	Not applicable
By result	Critical failure
	Major failure
	Minor failure

Given then that the word failure is specifically defined, then for a given application, quality and reliability and maintainability can be defined as follows.

Quality   Conformance to specification.

Reliability   The probability that an item will perform a required function, under stated conditions, for a stated period of time.

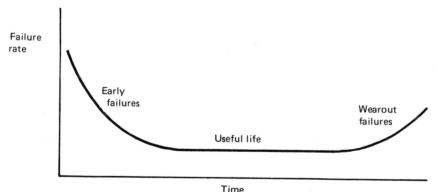

Figure 3.1

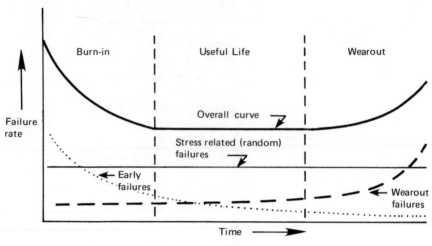

Figure 3.2   Bathtub curve

Table 3.2

	Known as	
Decreasing failure rate	Infant mortality Burn in Early failures	Usually related to manufacture and QA, e.g. welds, joints, connections, wraps, dirt, impurities, cracks, insulation or coating flaws, incorrect adjustment or positioning

Constant failure rate	Random failures
	Useful life
	Stress-related
	   failures
	Stochastic
	   failures
		Usually assumed to be stress-related failures. That is to say random fluctuations (transients) of stress exceeding the component strength (see Chapter 7). The design reliability referred to in figure 1.1 is of this type

Increasing failure rate	Wearout failures	Due to corrosion, oxidation, breakdown of insulation, atomic migration, friction wear, shrinkage, fatigue, etc.

Reliability is therefore the extension of quality into the time domain and may be paraphrased as 'the probability of non-failure in a given period'.

Maintainability   The probability that a failed item will be restored to operational effectiveness within a given period of time when the repair action is performed in accordance with prescribed procedures.

This, in turn, can be paraphrased as 'The probability of repair in a given time'.

3.2 FAILURE RATE AND MEAN TIME BETWEEN FAILURES

Requirements are seldom expressed by stating values of reliability or of maintainability. There are useful related parameters such as Failure Rate, Mean Time Between Failures and Mean Time to Repair which more easily describe them. In figure 3.2 the failure rate concept has been used to describe a variation of reliability with time. Figure 3.3 provides a model against which failure rate can be defined.

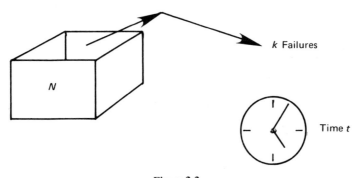

k Failures

N

Time *t*

Figure 3.3

The symbol for failure rate is λ (lambda). Assume a batch of N items and that at any time t a number k have failed. The cumulative time, T, will be Nt if it is assumed that each failure is replaced when it occurs. In a non-replacement case T is given by:

$$T = [t_1 + t_2 + t_3 \cdots t_k + (N - k)t]$$

where t_1 is the occurrence of the 1st failure, etc.

The Observed Failure Rate   This is defined: For a stated period in the life of an item, the ratio of the total number of failures to the total cumulative observed time. If λ is the failure rate of the N items then the observed λ is given by $\hat{\lambda} = k/T$ The $\frown$ (hat) symbol is very important since it indicates that k/T is only an estimate of λ. The true value will only be revealed when all N items have failed.

Making inferences about λ from values of k and T is the purpose of chapters 12 and 13. It should also be noted that the value of $\hat{\lambda}$ is the average over the period in question. The same value could be observed from increasing, constant and decreasing failure rates. This is analogous to the case of a motor car whose speed between two points is calculated as the ratio of distance to time although the velocity may have varied during this interval.

The Observed Mean Time Between Failures   This is defined: For a stated period in the life of an item the mean value of the length of time between consecutive failures, computed as the ratio of the total cumulative observed time to the total number of failures. If θ (theta) is the MTBF of the N items then the observed MTBF is given by $\hat{\theta} = T/k$. Once again the hat indicates a point estimate and the foregoing remarks apply. The use of T/k and k/T to define $\hat{\theta}$ and $\hat{\lambda}$ leads to the inference that $\theta = 1/\lambda$. This equality must be treated with caution since it is inappropriate to compute failure rate unless it is constant. It will be shown, in any case, that the equality is only valid under those circumstances. See section 3.5, equations (3.5) and (3.6).

The Observed Mean Time to Fail   This is defined: For a stated period in the life of an item the ratio of cumulative time to the total number of failures. Again this is T/k. The only difference between MTBF and MTTF is in their usage. MTTF is applied to items that are not repaired, such as components, and MTBF to items which are repaired. It must be remembered that the time between failures excludes the down time. MTBF is therefore mean UP time between failures. In figure 3.4 it is the average of the values of (t).

Mean Life   This is defined as the mean of the times to failure where each item is allowed to fail. This is often confused with MTBF and MTTF. It is important to understand the difference. MTBF and MTTF can be calculated over any period as, for example, confined to the constant failure rate portion of the Bathtub Curve (figure 3.2). Mean life on the other hand must include the failure of every item and therefore takes into account the wearout end of the curve. For constant failure rate situations only are they the same.

Failure rate, which has the unit of t^{-1}, is sometimes expressed as a percentage per 1000 h and sometimes as a number multiplied by a negative power of ten. As high reliability in components is now common 10^{-9} h^{-1} is often used. The

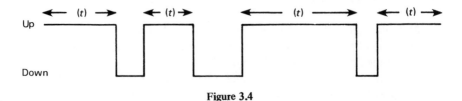

Figure 3.4

following three examples have the same value: 8500×10^{-9} h^{-1}; 8.5×10^{-6} h^{-1}; 0.85 % per 1000 h. Notice that these examples each have only two significant figures. It is seldom justified to exceed this level of accuracy particularly if failure rates are being used to carry out a reliability prediction — see chapter 15.

3.3 AVAILABILITY, DOWN TIME AND REPAIR TIME

In the first chapter mention was made of Availability as a useful parameter which describes the amount of available time. It is determined by both the reliability and the maintainability of the item. Returning to figure 3.4 it is the ratio of the (t) values to the total time. Availability is, therefore:

$$Av = \frac{\text{Up time}}{\text{Total time}}$$

$$= \frac{\text{Up time}}{\text{Up time} + \text{Down time}}$$

$$= \frac{\text{Average of } (t)}{\text{Average of } (t) + \text{Mean down time}}$$

$$= \frac{\text{MTBF}}{\text{MTBF} + \text{MDT}}$$

This is known as the steady-state availability and can be expressed as a ratio or as a percentage.

It is now necessary to consider Mean Down Time and Mean Time to Repair (MDT, MTTR). There is frequently confusion between the two and it is therefore important to understand the difference. Down time, or outage, is the period during which an equipment is in the failed state. A formal definition is usually avoided due to the difficulties of generalising about a parameter which may consist of different elements according to the system and its operating conditions. Consider the following examples which emphasise the problem:

(i) A system not in continuous use may develop a fault while it is idle. The fault condition may not become evident until the system is required for operation. Is down time to be measured from the incidence of the fault, from the start of an alarm condition, or from the time when the system would have been required?

(ii) In some cases it may be economical or essential to leave an equipment in a faulty condition until a particular moment or until several similar failures have accrued.

(iii) Repair may have been completed but it may not be safe to restore the system to its operating condition immediately. Alternatively due to a cyclic type of situation it may be necessary to delay. When does down time cease under these circumstances?

It is necessary, as can be seen from the above, to define down time as required for each system under given operating conditions and maintenance arrangements. MTTR and MDT, although overlapping, are not identical. Down time may commence before repair as in (i) above. Repair often involves an element of checkout or alignment which may extend beyond the outage. The definition and use of these terms will depend on whether availability or the maintenance resources are being considered.

The significance of these terms is not always the same depending upon whether a system, a replicated unit or a replacable module is being considered. The following comparison shows the effect of down time and repair time at different levels.

	Down time	Repair time
System	Determines availability and hence cost of lost revenue	Contributes to maintenance cost
Redundant unit	Determines system availability	Contributes to maintenance cost
Replaceable module	Influences spares level	Influences maintenance cost and availability of module as a spare part

Figure 3.5 shows the elements of down time and repair time.

(a) *Realisation Time*   This is the time which elapses before the fault condition becomes apparent. This element is pertinent to availability but does not constitute part of the repair time.

(b) *Access Time*   This involves the time, from realisation that a fault exists, to make contact with displays and test points and so commence fault finding. This does not include travel but the removal of covers and shields and the connection of test equipment. This is determined largely by mechanical design.

(c) *Diagnosis Time*   This is referred to as fault finding and includes adjustment of test equipment (e.g. setting up an oscilloscope or generator), carrying out checks (e.g. examining waveforms for comparison with a handbook), interpretation of information gained (this may be aided by algorithms), verifying the conclusions drawn and deciding upon the corrective action.

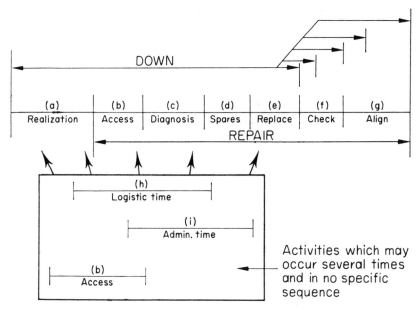

Figure 3.5   Elements of down time and repair time

(d) *Spare Part Procurement*   Part Procurement can be from the 'tool box', by cannibalisation or by taking a redundant identical assembly from some other part of the system. The time taken to move parts from a depot or store to the system is not included, being part of the logistic time.

(e) *Replacement Time*   This involves removal of the faulty LRA (Least Replaceable Assembly) followed by connection and wiring, as appropriate, of a replacement. The LRA is the replaceable item beyond which fault diagnosis does not continue. Replacement time is largely dependent on the choice of LRA and on mechanical design features such as the choice of connectors.

(f) *Checkout Time*   This involves verifying that the fault condition no longer exists and that the system is operational. It may be possible to restore the system to operation before completing the checkout in which case, although a repair activity, it does not all constitute down time.

(g) *Alignment Time*   As a result of inserting a new module into the system adjustments may be required. As in the case of checkout some or all of the alignment may fall outside the down time.

(h) *Logistic Time*   This is the time consumed waiting for spares, test gear, additional tools and manpower to be transported to the system.

(i) *Administrative Time*   This is a function of the system user's organisation. Typical activities involve failure reporting (where this affects down time), allocation of repair tasks, manpower changeover due to demarcation arrangements, official breaks, disputes, etc.

Activities (2)–(7) are called Active Repair Elements and activities (8) and (9) Passive Repair Activities. Realsiation time is not a repair activity but may be included in the MTTR where down time is the consideration. Checkout and alignment, although utilising manpower, can fall outside the down time. The Active Repair Elements are determined by design, maintenance arrangements, environment, manpower, instructions, tools and test equipment. Logistic and Administrative time is mainly determined by the maintenance environment, that is to say the location of spares, equipment and manpower and the procedures for allocating tasks.

Another parameter related to outage is Repair rate (μ). It is simply the down time expressed as a rate, therefore:

$$\mu = 1/\text{MTTR}$$

Equipment and components are often described by means of the word Life. This must not be confused with MTBF or MTTF. It refers to the onset of wearout (see figure 3.2) and not to the level of reliability during the useful life. An item may be short lived but highly reliable as, for example, a magnetron or have long life with only medium reliability as is the case with some resistors.

3.4 CHOOSING THE APPROPRIATE PARAMETER

It is clear that there are many parameters available for describing the reliability and maintainability characteristics of an item. In any particular instance there is likely to be a parameter more appropriate than the others. Although there are no hard and fast rules the following guidelines may be of some assistance:

Failure Rate	Only relevant when constant. Applicable to electronic components.
MTBF and MTTF	Often used to describe equipment or system reliability. Of use when calculating maintenance costs.
Reliability	Used where the probability of failure is of interest as, for example, in aircraft landings where safety is the prime consideration.
Maintainability	Seldom used as such.
Mean Time To Repair	Often expressed in percentile terms such as, the 95 percentile repair time shall be 1 hour. This means that only 5 per cent of the repair actions shall exceed 1 h.
Mean Down Time	Used where the outage affects system reliability or availability. Often expressed in percentile terms.

Availability   Very useful where there is a high cost of lost revenue
                                due to outage. Combines reliability and
                                maintainability.

There are sources of standard definitions such as:

> British Standard BS 4200 (Part 1)
> IEC Publication 271
> US MIL STD 721B
> UK Defense Standard 00-5 (Part 1)

It is, however, not desirable to use standard sources of definitions in such a way
as to avoid specifying the terms which are to be used in a specification or
contract. It is all too easy to 'define' the terms by calling up one of the foregoing
standards. It is far more important that terms are fully understood before they
are used and if this is achieved by defining them for specific situations then so
much the better. The danger in specifying that all terms shall be defined by a
given published standard is that each person assumes that he knows the meaning
of each term and these are not read or discussed until a dispute arises. The most
important area involving definition of terms is that of contractual involvement
where mutual agreement as to the meaning of terms is essential. Chapter 18 will
emphasise the dangers of ambiguity.

3.5 INTERRELATIONSHIP OF TERMS

Returning to the model in figure 3.3 consider the probability of an item failing
in the interval between t and $t + dt$. This can be described in two ways:

The probability of failure in the interval t to $t + dt$ given that it has survived
until time t. This is

$$\lambda(t)\, dt$$

where $\lambda(t)$ is the FAILURE RATE.

The probability of failure in the interval t to $t + dt$ unconditionally. This is

$$f(t)\, dt$$

where $f(t)$ is the FAILURE PROBABILITY DENSITY FUNCTION.

The probability of survival to time t is defined as the reliability, $R(t)$. The rule of
conditional probability therefore dictates that:

$$\lambda(t)\, dt = \frac{f(t)\, dt}{R(t)}$$

Therefore

$$\lambda(t) = \frac{f(t)}{R(t)} \tag{3.1}$$

However, if $f(t)$ is the probability of failure in dt then:

$$\int_0^t f(t)\, dt = \text{probability of failure 0 to } t = 1 - R(t)$$

Differentiating both sides:

$$f(t) = -\frac{dR(t)}{dt} \tag{3.2}$$

Substituting (3.2) into (3.1)

$$-\lambda(t) = \frac{dR(t)}{dt} \cdot \frac{1}{R(t)}$$

Therefore integrating both sides:

$$-\int_0^t \lambda(t)\, dt = \int_1^{R(t)} dR(t)/R(t)$$

A word of explanation concerning the limits of integration is required. $\lambda(t)$ is integrated with respect to time from 0 to t. $1/R(t)$ is, however, being integrated with respect to $R(t)$. Now when $t = 0, R(t) = 1$ and at t the reliability $R(t)$ is, by definition, $R(t)$. Integrating then:

$$-\int_0^t \lambda(t)\, dt = \log_e R(t) \Big|_1^{R(t)}$$

$$= \log_e R(t) - \log_e 1$$

$$= \log_e R(t)$$

But if $a = e^b$ then $b = \log_e a$, so that:

$$R(t) = \exp\left[-\int_0^t \lambda(t)\, dt\right] \tag{3.3}$$

If failure rate is now assumed to be constant:

$$R(t) = \exp\left[-\int_0^t \lambda\, dt\right] = \exp -\lambda t \Big|_0^t \tag{3.4}$$

Therefore $R(t) = e^{-\lambda t}$

In order to find the MTBF consider figure 3.3 again. Let $N - k$, the number surviving at t, be $N_s(t)$. Then $R(t) = N_s(t)/N$.

In each interval dt the time accumulated will be $N_s(t)\,dt$.

At ∞ the total will be $\displaystyle\int_0^\infty N_s(t)\,dt$.

Hence the MTBF will be given by:

$$\theta = \int_0^\infty \frac{N_s(t)\,dt}{N} = \int_0^\infty R(t)\,dt$$

$$\theta = \int_0^\infty R(t)\,dt \tag{3.5}$$

This is the general expression for MTBF and always holds. In the special case of $R(t) = e^{-\lambda t}$ then

$$\theta = \int_0^\infty e^{-\lambda t}\,dt$$

$$\theta = \frac{1}{\lambda} \quad ** \tag{3.6}$$

Note
1. If failure rate is constant and, hence, $R = e^{-\lambda t} = e^{-t/\theta}$ then after one MTBF the probability of survival, $R(t)$ is e^{-1} which is 0.37.
2. If t is small $e^{-\lambda t}$ approaches $1 - \lambda t$. For example if $\lambda = 10^{-5}$ and $t = 10$ then $e^{-\lambda t}$ approaches $1 - 10^{-4} = 0.9999$.
3. Since $\theta = \int_0^\infty R(t)\,dt$ it is useful to remember that $\int_0^\infty A e^{-B\lambda t} = \dfrac{A}{B\lambda}$
4. ** Used as the reciprocal of failure rate MTBF is an alternative parameter for describing the random failure portion of the bathtub curve.

Part II

Achieving Reliability and Maintainability Objectives

4 Design and Assurance for Reliability and Maintainability

4.1 INHERENT DESIGN LEVELS

It is essential to believe that the design of an item establishes its potential reliability and maintainability. It is a fact of life that the transition from drawings to hardware always results in achieved levels lower than the original design objectives. It is therefore necessary to design to specified levels of reliability and maintainability higher than the field requirements and to follow up with assurance activities aimed at minimising the failures which arise during manufacture and use. Figure 1.1 illustrates the concept of a reliability level fixed by the 'drawings' and a lower field level due to the many failure possibilities arising from manufacture and subsequent use. The author makes no apology for repetition since this is fundamental to Reliability Engineering. This chapter will outline the activities and techniques which can be used both in design, to develop equipment which can meet the R and M objectives and also in manufacture and use, to minimise failures.

4.2 ACTIVITIES IN DESIGN

4.2.1 Specifying and Allocating the Requirement

The prime objective of a reliability and maintainability programme is to assure adequate performance consistent with minimal maintenance costs. This can only be achieved if, in the first place, objectives are set and described by suitable parameters. The intended use and environment of a system must be accurately appraised in order to set realistic objectives and, in the case of contract design, the customer requirements must be delineated. It may well be that the customer has not considered these points and guidance may therefore be necessary in persuading him to set reasonable targets with regard to the technology, environment and overall cost envisaged. Appropriate parameters have then to be chosen as already discussed.

System reliability and maintainability will be specified, perhaps in terms of MTBF and MTTR, and values have then to be assigned to each separate unit. Thought must be given to the ALLOCATION of these values throughout the system such that the overall objective is achieved without overspecifying the

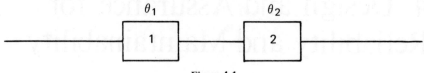

θ_1 θ_2

Figure 4.1

requirement for one unit whilst underspecifying for another. Figure 4.1 shows a
simple system comprising two units connected in such a way that neither may
fail if the system is to perform. We shall see in chapter 15 that the system MTBF
is given by:

$$\theta_s = \frac{\theta_1\theta_2}{\theta_1 + \theta_2}$$

If the design objective for θ_s is 1000 h then this may be met by setting θ_1 and
θ_2 both at 2000 h. An initial analysis of the two units, however, could reveal
that unit 1 is twice as complex as, and hence likely to have half the MTBF of,
unit 2. If the reliability is allocated equally as suggested then the design task will
be comparatively easy for unit 2 and unreasonably difficult for unit 1. Ideally
the allocation of MTBF should be weighted so that:

$$2\theta_1 = \theta_2$$

Hence
$$\theta_s = \frac{2\theta_1^2}{3\theta_1} = \frac{2\theta_1}{3} = 1000 \text{ h}$$

Therefore
$$\theta_1 = 1500 \text{ h}$$

and
$$\theta_2 = 3000 \text{ h}$$

In this way the overall objective is achieved with the optimum design requirement
being placed on each unit. The same philosophy should be applied to the
allocation of repair times such that more attention is given to repair times in the
high failure rate areas.

System reliability and maintainability will not necessarily be defined by a
single MTBF and MTTR. Failures are classified, chapter 3 (section 3.1), under
many headings and it is usually appropriate to specify a number of different
values against various types of failure. For example the requirement for an item
of communications equipment might be stated as follows:

Complete Failure — Failure resulting from a deviation in characteristics such as
to cause a complete lack of the required function.
MTBF — 10 000 h   MTTR — 2 h

Partial Failure — Failure resulting from a deviation in characteristics beyond
specified limits but not such as to cause complete lack of the required
function.
MTBF — 5000 h   MTTR — 8 h

4.2.2 Applying Design Techniques

A. Derating   The principle of operating a component part below the rated stress
level of a parameter in order to obtain a longer or more reliable life is well
known. It is of particular interest in electronics where underrating of voltage and
temperature produce spectacular improvements in reliability. Stresses can be
divided into two broad categories — environmental and operating.

Operating stresses are present when a device is active. Examples are voltage,
current, self-generated temperature and self-induced vibration. These have a
marked effect on the frequency of random failures as well as hastening wearout.
Figure 4.2 shows the relationship of failure rate to voltage and temperature stress
for a typical wet aluminium capacitor.

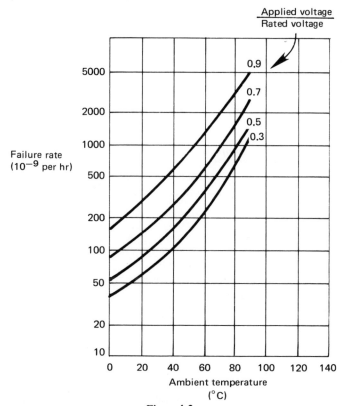

Figure 4.2

Notice that a 5 to 1 improvement in failure rate is obtained by either a reduction in voltage stress from 0.9 to 0.3 or a 30 °C reduction in temperature. The relationship of failure rate to stress in electronic components is often described by a form of the Arrhenius equation which relates chemical reaction rate to temperature. Applied to random failure rate the following two forms are often used:

$$\lambda_2 = \lambda_1 \ \exp K \left(\frac{1}{T_1} - \frac{1}{T_2} \right)$$

$$\lambda_2 = \lambda_1 \left(\frac{V_2}{V_1} \right)^n G^{(T_2 - T_1)}$$

V_2, V_1, T_2 and T_1 are voltage and temperature levels. λ_2 and λ_1 are failure rates at those levels. K, G and n are constants.

It is dangerous to use such empirical formulae outside the range over which they have been substantiated. Unpredicted physical or chemical effects may occur which render them inappropriate and the results, therefore, can be misleading. Mechanically the principle of excess material is sometimes applied to increase the strength of an item. It must be remembered that this can sometimes have the reverse effect and the elimination of large sections in a structure can increase the strength and hence reliability.

B. Environmental Stress Protection   Environmental stress hastens the onset of wearout by contributing to physical deterioration. Included are:

Stress	*Symptom*	*Action*
High temperature	Insulation materials deteriorate. Chemical reactions accelerate	Dissipate heat. Minimise thermal contact. Use fins. Increase conductor sizes on PCBs. Provide conduction paths
Low temperature	Mechanical contraction damage. Insulation materials deteriorate	Apply heat and thermal insulation
Thermal shock	Mechanical damage within LSI components	Shielding
Mechanical shock	Component and connector damage	Mechanical design. Use of mountings
Vibration	Hastens wearout and causes connector failure	Mechanical design

Humidity	Coupled with temperature cycling causes 'pumping' — filling up with water	Sealing. Use of silica gel
Salt atmosphere	Corrosion and insulation degradation	Mechanical protection
Electromagnetic radiation	Interference to electrical signals	Shielding and part selection
Dust	Long-term degradation of insulation. Increased contact resistance	Sealing. Self-cleaning contacts
Biological effects	Decayed insulation material	Mechanical and chemical protection
Acoustic noise	Electrical interference due to microphonic effects	Mechanical buffers
Reactive gases	Corrosion of contacts	Physical seals

C. Reduction of Complexity   Higher scales of integration in electronic technology enable circuit functions previously requiring many hundreds, or thousands, of devices to be performed by a single component. Hardware failure is restricted to either the device or its connections (sometimes 40 pins) to the remaining circuitry. A reduction in total device population and quantity leads, in general, to higher reliability.

Standard circuit configurations help to minimise component populations and allow the use of proven reliable circuits. Regular planned design reviews provide an opportunity to assess the economy of circuitry for the intended function. Digital circuits provide an opportunity for reduction in complexity by means of logical manipulation of the expressions involved. This enables fewer logic functions to be used in order to provide a given result.

D. Part Selection   Since hardware reliability is largely determined by the component parts their reliability and fitness for purpose cannot be overemphasised. The choice often arises between standard parts with proven performance which just meet the requirement and special parts which are totally applicable but unproven. Consideration of design support services when selecting a component source may be of prime importance when the application is a new design. General considerations should be:

Function needed and the environment in which it is to be used.
Critical aspects of the part as, for example, limited life, procurement time, contribution to overall failure rate, cost, etc.
Availability — number of different sources.

Stress — given the application of the component the stresses applied to it and the expected failure rate. The effect of burn in and screening on actual performance.

E. Redundancy   This involves the use of active additional units or of standby units. Reliability may be enahnced by this technique which can be applied in a variety of configurations:

Active Redundancy — Full — With duplicated units, all operating, one surviving unit ensures non-failure.
    Partial — A specified number of the units may fail as, for example, 2 out of 4 engines on an aircraft. Majority voting systems often fall in this category.
    Conditional — A form of redundancy which occurs according to the failure mode.
Standby Redundancy — Involves extra units which are not brought into use until the failure of the main unit is sensed.
Load Sharing — Active redundancy where the failure of one unit places a greater stress on the remaining units.
Redundancy and Repair — Where redundant units are subject to immediate or periodic repair the system reliability is influenced both by the unit reliability and the repair times.

The decision to use redundancy must be based on an analysis of the trade-offs involved. It may prove the only available method when other techniques have been exhausted. Its application is not without penalties since it increases weight, space and cost and the increase in number of parts results in an increase in maintenance and spares holding costs. In general the reliability gain obtained from additional elements decreases beyond a few duplicated elements due to the series reliability of switching or other devices needed to implement the particular configuration employed. Chapter 15 deals, in detail, with the quantitative effects of redundancy.

F. The Choice of Technology   There are often alternative solutions available which equally satisfy the design requirement. The choice is not always straightforward since such alternatives have to be compared in terms of present cost, future cost (as the technology and market develops), reliability, ease of repair, spares availability, size, power, weight and so on. Examples are:

Electro-mechanical *v*. Solid State
    Encapsulated reed relays replaced by solid state switching.
    Increased switch reliability against a more complex control function.
Electrostatic *v*. Electrosensitive
    Plain paper reprographics by means of attracting powder to a charged surface instead of direct thermal or electrolytic marking of an impregnated paper.

Convenience of using plain paper against the mechanical complexity of the electrostatic charging assembly.

Light Emitting Diode *v*. Filament Device

Convenient and reliable indicators and alpha numeric displays.

Low power and cost against low output intensity.

Discrete Electronic Components *v*. Integrated Circuits

Simplified circuit design using high package density.

Power, space and weight advantage v new maintenance and test skills.

Alternative Materials

Nylon and other polymer based materials are often used for machined parts in place of traditional metals.

Reliability and size *v*. diagnosis and test equipment costs.

G. Failure Analysis and Prediction of R and M   It must be stressed that failure analysis and reliability prediction do not, of themselves, contribute to reliability or maintainability. They are merely tools which can be used to pinpoint those actions which will cause improvements. Failure mode analysis allows maintainability to be evaluated at the same time as reliability by giving the opportunity for each failure type to be examined for its ease of diagnosis. Chapter 7 will elaborate. These activities will lead to trade-offs between the different methods of achieving R and M. The reliability of a wrapped joint might be traded for the maintainability associated with a connector.

4.2.3 Qualification Testing

The purpose of Qualification Testing is to ensure that a product meets all the requirements laid down in the Engineering Specification. This should not be confused with product testing which takes place after manufacture. Items to be verified are:

Function — Specified performance at defined limits and margins.

Environment — Ambient Temperature and Humidity for use, storage, etc.

Performance at the extremes of the specified environment should be included.

Life — At specified performance levels and under storage conditions.

Reliability — Observed MTBF under all conditions.

Maintainability — MTTR for defined test equipment, spares, manual and staff.

Maintenance — Is the routine and corrective maintenance requirement compatible with use?

Packaging and Transport — Test under real conditions including shock tests.

Physical characteristics — Size, weight, power consumption, etc.

Ergonomics — Consider interface with operators and maintenance personnel.

Testability — Consider test equipment and time required for production models.

Safety — Use an approved test house such as BSI or the British Electrotechnical Approvals Board.

4.3 ASSURANCE ACTIVITIES

4.3.1 Burn In

The early (infant mortality) failures shown by the decreasing failure rate curve of the bathtub diagram in figure 3.2 are frequently induced during testing by a 'burn in' period. This can take between 20 and 500 h, is determined by field experience, and is intended to bring the product into the random failures portion of the bathtub prior to commissioning. It may be conducted in an environment similar to that anticipated for field use or a more severe environment intended to accelerate early, random and wearout failures. A method of analysis which determines whether the failure rate is reducing or random will be explained in chapter 13.

4.3.2 Preventive Maintenance

A. Routine   This consists of cleaning, adjusting and lubricating on a planned basis. It is difficult to quantify the effect of this type of activity except over a long period therefore opinion is divided as to its value. Examples are often quoted where more failures are caused than prevented. A common-sense approach should be taken in defining these routines and they should be evaluated on early models in the field.

B. Preventive Replacement   This requires a knowledge of the wearout distribution of a particular item in the equipment. A policy of replacing items at a particular 'age' even if they have not failed is adopted. This is only applicable to the wearout situation and figure 4.3 gives the example of the failure distribution of a filament lamp. In this example the failures are distributed according to the Gaussian (normal) law. Other examples may exhibit more complex distributions but the principle of selecting a time for automatic replacement based on the probability of failure is the same. In this example if lamps are replaced at 1000 h then 50 per cent will fail before replacement. If it is decided, to replace them earlier, at a time one standard deviation before the mean, then approximately 15.9 per cent will fail before replacement. Some potential useful life is thus traded against the inconvenience and cost of a failure.

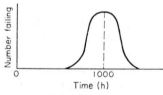

Figure 4.3

C. Dormant Failures   Many faults do not cause immediate system failure and hence cause no alarm. Only a further fault or combination of stress conditions reveals this condition, nevertheless its existance may well accelerate a system failure. A failed component in a circuit which, although not causing malfunction, places greater stresses on those remaining is one example. Another is a failed alarm circuit, being particularly undesirable since it may conceal a further failure until damage has resulted. Routine checks for dormant faults are expensive but may well be justified as part of the routine maintenance procedure for key items.

D. Drift Conditions (Degradation)   Measurements of key parameters aimed at detecting whether they are drifting toward the limit would result in preventive replacement and enhanced reliability. This is an expensive procedure and is only likely to be justified in a few cases.

E. Repair and Redundancy   Periodic inspection and early repair of unattended redundant units also enhances reliability. MTBF is thus dependent on repair times as well as unit reliability.

4.3.3 Quality Control and Test

The early failures of the bathtub curve are associated both with manufacturing processes (soldering, wrapping, assembly) and with marginally acceptable components. Both types of failure will be reduced by the application of quality control particularly if inspection and test are fed with failure data derived from field information. The procedures and methods involved in a QA system would easily occupy another volume and there is already much excellent literature on the subject. Field feedback is an activity specifically related to R and M and chapter 11 is therefore devoted to the topic.

5 Design Factors Influencing Down Time

The two main factors governing down time are equipment design and maintenance philosophy. In general it is the active repair elements which are determined by the design and the passive elements which are governed by the maintenance philosophy. The designer must be aware of the maintenance environment and of the possible equipment failure modes. He must understand that production difficulties all too often become field problems since, if assembly is difficult, maintenance will be well nigh impossible. Achieving acceptable repair times involves facilitating diagnosis and repair. The main design parameters are as follows.

5.1 ACCESS

Low-reliability parts should be the most accessible and must be easily removable with the minimum of disturbance. There must be adequate room to withdraw such devices without striking or damaging other parts. On the other hand the technician must be discouraged from removing and checking easily exchanged items as a substitute for the correct diagnostic procedure. The use of such techniques as captive screws and fasteners is highly desirable as they are faster to use and eliminate the risk of losing screws in the equipment. Standard fasteners and covers become familiar and hence easier to use. The use of outriggers, which enables printed boards to be tested whilst still electrically connected to the system, can help to reduce diagnosis time. This type of on-line diagnosis can induce faults and is sometimes discouraged. In general it is a good thing to minimise on-line testing by employing easily interchanged units together with alarms and displays providing diagnostic information and easy identification of the faulty unit.

Every LRA (Least Replaceable Assembly) should be capable of removal without the removal of any other LRA or part. The size of the LRA affects the speed of access. The overall aim is for speedy access consistent with minimum risk of accidental damage.

5.2 ADJUSTMENT

The amount of adjustment required during normal system operation, and after

LRA replacement, can be minimised (or eliminated) by generous tolerancing in the design aimed at low sensitivity to drift.

Where adjustment is by a screwdriver or other tool care should be taken to ensure that damage cannot be done to the equipment. Guide holes, for example, can prevent a screwdriver from slipping.

Where adjustment requires that measurements are made, or indicators observed, then the displays or meters should be easily visible whilst the adjustment is made.

It is usually necessary for adjustments and alignments to be carried out in a sequence and this must be specified in the maintenance instructions. The designer should understand that where drift in a particular component can be compensated for by the adjustment of some other item then, if that adjustment is difficult or critical, the service engineer will often change the drifting item regardless of its cost.

5.3 BUILT-IN TEST EQUIPMENT

As with all test equipment, built-in test equipment (BITE) should be an order of magnitude more reliable than the system of which it is part in order to minimise the incidence of false alarms or incorrect diagnosis. Poor reliability BITE will grossly reduce the system effectiveness.

The number of connections between the system and the built-in test equipment should be minimised to reduce the probability of system faults induced by them. It carries the disadvantages of being costly, inflexible (designed around the system it is difficult to modify) and of requiring some means of self-checking. In addition it carries a weight, volume and power supply penalty but, on the other hand, greatly reduces the time required for realisation, diagnosis and checkout.

5.4 CIRCUIT LAYOUT AND HARDWARE PARTITIONING

It is not too early to consider maintainability when designing and laying out circuitry. In some cases it is possible to identify a logical sequence of events or signal flow through a circuit and fault diagnosis is helped by a component layout which reflects this logic. Components should not be so close together as to make damage likely when removing and replacing a faulty item.

The use of integrated circuits introduces difficulties. Their small size and large number of leads makes it necessary for connections to be small and close together which increases the possibility of damage during maintenance. In any case field maintenance at circuit level is almost impossible due to the high function density involved. Due to the high maintenance cost of removing and resoldering these devices the question of plug in ICs arises. Another point of view emphasises that IC sockets increase both cost and the possibility of connector failure. The decision for or against is economic and must be made on the basis of field failure rate, socket cost and repair time. The IC is a functional unit in itself and therefore circuit layout is less capable of representing the circuit function.

On a higher level, hardware partitioning involves grouping units and modules functionally and physically as an aid to diagnosis.

5.5 CONNECTIONS

Connections present a classic trade-off between reliability and maintainability. The following types of connection are ranked in order of reliability starting with the most reliable. A comparison of failure rates is made by means of the following:

Wrapped joint	0.05×10^{-9} per h
Welded connection	0.3×10^{-9} per h
Machine-soldered joint	0.4×10^{-9} per h
Crimped joint	0.5×10^{-9} per h
Hand-soldered joint	1.0×10^{-9} per h
Edge connector (per pin)	2.0×10^{-9} per h

Since edge connectors are less reliable than soldered joints there is a balance between having a few large plug-in units and a larger number of smaller throw-away units with the attendant reliability problem of additional edge connectors. Boards terminated with wrapped joints rather than with edge connectors are an order more reliable from the point of view of the connections, but the maintainability penalty can easily outweigh the reliability advantage. Consider the time taken to make ten or twenty wrapped joints compared with the time taken to plug in a board equipped with edge connectors.

The following approximate times for making the different types of connection assume the appropriate tools are available:

Edge connector (multi-contact)	10 s
Solder joint (single-wire)	20 s
Wrapped joint	50 s

As can be seen the maintainability ranks in the opposite order to reliability. In general a high-reliability connection is required within the LRA where maintainability is a secondary consideration. The interface between the LRA and the system requires a high degree of maintainability and the plug-in or edge connector is justified. If the LRA is highly reliable, and therefore unlikely to require frequent replacement, termination by the reliable wrapped joints could be justified. On the other hand a medium or low reliability unit would require plug and socket connection for quick interchange.

The reliability of a solder joint, hand or flow, is extremely sensitive to the quality control of the manufacturing process.

Where cable connectors are used it should be ensured, by labelling or polarising, that plugs will not be wrongly inserted in sockets or inserted in wrong sockets. Mechanical design should prevent insertion of plugs in the wrong configuration and also prevent damage to pins by clumsy insertion.

Where several connections are to be made within or between units the complex of wiring is often provided by means of a cableform (loom) and the terminations (plug, solder or wrap) made according to an appropriate document. The cableform should be regarded as an LRA and local repairs should not be attempted. A faulty wire may be cut back, but left in place, and a single wire added to replace the link providing that this does not involve the possibility of electrical pickup or misphasing.

5.6 DISPLAYS AND INDICATORS

Displays and indicators may be effective in reducing diagnostic, checkout and alignment elements of active repair time. Simplicity should be the keynote and a 'go, no go' type of meter or display will require only a glance. The use of stark colour changes, or other obvious means, to divide a scale into areas of 'satisfactory operation' and 'alarm' should be used. Sometimes a meter, together with a multiway switch, is used to monitor several parameters in a system. It is desirable that the anticipated (normal) indication be the same for all the applications of the meter so that the correct condition is shown by little or no movement as the instrument is switched to the various test points. Displays should never be positioned where it is difficult, dangerous or uncomfortable to read them.

For an alarm condition an audible signal, as well as visual displays, is needed to draw attention to the fault. Displays in general, and those relating to alarm conditions in particular, must be more reliable than the parent system since a failure to indicate an alarm condition is potentially dangerous.

If an equipment is unattended then some alarms and displays may have to be extended to another location and the reliability of the communications link then becomes important to the availability of the system.

The following points concerning meters are worth noting:

(i)   False readings can result from parallax effects due to scale and pointer being in different planes. A mirror behind the pointer helps to overcome this difficulty.

(ii)   Where a range exists outside which some parameter is unacceptable then either the acceptable or the unacceptable range should be coloured or otherwise made readily distinguishable from the rest of the scale — figure 5.1a.

(iii)   Where a meter displays a parameter which should normally have a single value then a centre-zero instrument can be used to advantage and the

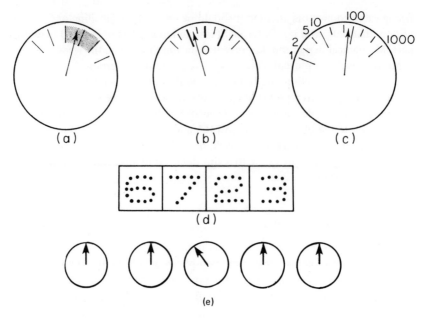

Figure 5.1   Meter displays. (a) Scale with shaded range, (b) scale with limits, (c) logarithmic scale, (d) digital display, (e) alignment of norms

circuitry configured such that the normal acceptable range of values falls within the mid-zone of the scale — figure 5.1b.

(iv) Linear scales are easier to read and less ambiguous than logarithmic scales and consistency in the choice of scales and ranges minimises the possibility of misreading — figure 5.1c. On the other hand there are occasions when the use of a non-linear response or false-zero meter is desirable.

(v) Digital displays are now widely used and are superior to the analogue pointer-type of instrument where a reading has to be recorded — figure 5.1d. The analogue type of display is preferable when a check or adjustment within a range is required.

(vi) When a number of meters are grouped together it is desirable that the pointer positions for the NORMAL condition are alike. Figure 5.1e shows how easily an incorrect reading is noticed.

Consistency in the use of colour codes, symbols and labels associated with displays is highly desirable. Filament lamps are not particularly reliable and should be derated. More reliable LEDs and liquid crystal displays are now widely used.

All displays should be positioned as near as possible to the location of the function or parameter to which they refer and should be mounted in an order relating to the sequence of adjustment. Unnecessary displays merely complicate the maintenance task and do more harm than good. Meters need be no more accurate than the measurement requirement of the parameter involved.

5.7 HANDLING, HUMAN AND ERGONOMIC FACTORS

Major handling points to watch are:

Weight, size and shape of removable modules. The LRA should not be capable of self-damage due to its own instability as in the case of a thin lamina construction.

Protection of sharp edges and high-voltage sources. Even an unplugged module may hold dangerous charges on capacitors.

Correct handles and grips reduce the temptation to use components for that purpose.

When an inductive circuit is broken by the removal of a unit then the earth return should not be via the frame. A separate earth return via a pin or connection from the unit should be used.

The following ergonomic factors also influence active repair time:

Design for minimum maintenance skills considering what type of personnel are actually available.

Beware of overminiaturisation — incidental damage is more likely.

Consider comfort and safety of personnel when designing for access; e.g. body position, movements, limits of reach and span, limit of strength in various positions, etc.

Illumination — fixed and portable.

Shield from environment (weather, damp, etc.) and from stresses generated by the equipment (heat, vibration, noise, gases, moving parts, etc.) since repair is slowed down if the maintenance engineer has to combat these factors.

5.8 IDENTIFICATION

Identification of components, test points, terminals, leads, connectors and modules is helped by standardisation of appearance. Colour codes should not be complex since over 5 per cent of the male population suffer from some form of colour blindness. Simple, unambiguous, numbers and symbols help in the identification of particular functional modules. The physical grouping of functions simplifies the signs required to identify a particular circuit or LRA.

5.9 INTERCHANGEABILITY

Where LRAs are interchangeable this simplifies diagnosis, replacement and checkout due to the element of standardisation involved. Spares provisioning then becomes slightly less critical in view of the possibility of using a non-essential, redundant, unit to effect a repair in some other part of the system. Cannibalisation of several failed LRAs to yield a working module also becomes possible although this should never become standard field practice.

The smaller and less complex the LRA the greater the possibility of standardisation and hence interchangeability. The penalty lies in the number of interconnections, between LRAs and the system (less reliability) and the fact that the diagnosis is referred to a lower level (greater skill and more equipment).

Interchange of non-identical boards or units should be made mechanically impossible. At least, pin conventions should be such that insertion of an incorrect board cannot cause damage either to that board or to other parts of the equipment. Each value of power supply must always occupy the same pin number.

5.10 LEAST REPLACEABLE ASSEMBLY

The LRA is that replaceable module at which local fault diagnosis ceases and direct replacement occurs. Failures are traced only to the LRA, which should be easily removable (see section 5.5), replacement LRAs being the spares holding. It should rarely be necessary to remove an LRA in order to prove that is faulty, and no LRA should require the removal of any other LRA for diagnosis or for replacement.

The choice of level of the LRA is one of the most powerful factors in determining maintainability. The larger the LRA the faster the diagnosis. Maintainability, however, is not the only factor in the choice of LRA. As the size of the LRA increases so does its cost and the cost of spares holding. The more expensive the LRA the less likely is a throwaway policy to be applicable. Also a larger LRA is less likely to be interchangeable with any other. The following compares various factors as the size of LRA increases:

System maintainability	Improves
LRA reliability	Decreases
Cost of system testing (equipment and manpower)	Decreases
Cost of individual spares	Increases
Number of types of spares	Decreases

5.11 MOUNTING

If components are mounted so as to be self-locating then replacement is made easier. Mechanical design and layout of mounting pins and brackets can be made to prevent transposition where this is undesirable as in the case of a transformer which must not be connected the wrong way round. Fragile components should be mounted as far as possible from handles and grips.

5.12 COMPONENT PART SELECTION

Main factors effecting repair times are:

Availability of spares — delivery.
Reliability/deterioration under storage conditions.
Ease of recognition.
Ease of handling.
Cost of parts.
Physical strength and ease of adjustment.

5.13 REDUNDANCY

Circuit redundancy within the LRA (usually unmonitored) increases the
reliability of the module, and this technique can be used in order to make it
sufficiently reliable to be regarded as a throwaway unit.

Redundancy at the LRA level permits redundant units to be removed for
preventive maintenance whilst the system remains in service.

Although improving both reliability and maintainability redundant units
require more space and weight. Capital cost is increased and the additional units
require more spares and generate more maintenance. System availability is thus
improved but both preventive and corrective maintenance costs increase with the
number of units.

5.14 SAFETY

Apart from legal and ethical considerations safety hazards increase active repair
time by requiring greater care and attention. An unsafe design will encourage
short cuts or the omission of essential activities. Accidents add, very substantially,
to the repair time.

Where redundancy exists routine maintenance can be carried out after
isolation of the unit from high voltage and other hazards. In some cases routine
maintenance is carried out under power in which case appropriate safeguards
must be incorporated in the design. The following practices should be the norm:

Isolate high voltages under the control of microswitches which are automatically
    operated during access. The use of a positive interlock should bar access unless
    the condition is safe.
Weights should not have to be lifted or supported.
Use appropriate handles.
Provide physical shielding from high voltage, high temperature, etc.
Eliminate sharp points and edges.
Install alarm arrangements. The exposure of a distinguishing colour when safety
    covers have been removed is good practice.
Ensure adequate lighting.

5.15 SOFTWARE

The availability of programmable LSI (large scale integration) devices has revolutionised the approach to circuit design. More and more electronic circuitry is being replaced by a standard microprocessor architecture with the individual circuit requirements achieved within the software (program) which is held in the memory section of the hardware. Under these conditions diagnosis can no longer be supported by circuit descriptions and measurement information. Complex sequences of digital processing make diagnosis impossible with traditional test equipment.

Production testing of this type of printed board assembly is only possible with sophisticated computer driven automatic test equipment (ATE) and, as a result, field diagnosis can only be to board level. Where printed boards are inter-connected by data highways carrying dynamic digital information even this level of fault isolation may require field test equipment consisting of a microprocessor loaded with appropriate software for the unit under repair.

5.16 STANDARDISATION

Standardisation leads to improved familiarisation and hence shorter repair times. The number of different tools and test equipment is reduced as is the possibility of delay due to having incorrect test gear. Fewer types of spares are required reducing the probability of exhausting the stock.

5.17 TEST POINTS

Test points are an interface between test equipment and the system and are for the purpose of diagnosis, adjustment, checkout, calibration and monitoring for drift. Their provision is largely governed by the level of LRA chosen and they will usually not extend beyond what is necessary to establish that an LRA is faulty. Test points within the LRA will be dictated by the type of board test carried out in production or in second line repair.

In order to minimise faults caused during maintenance, test points should be accessible without the removal of covers and should be electrically buffered to protect the system from misuse of test equipment. Standard positioning also reduces the probability of incorrect diagnosis resulting from wrong connections. Test points should be grouped in such a way as to facilitate sequential checks. The total number should be kept to a minimum consistent with the diagnosis requirements. Unnecessary test points are likely to reduce rather than increase maintainability.

The above 17 design parameters relate to the product and not to the maintenance philosophy. Their main influence is on the active repair elements such as diagnosis, replacement, checkout, access and alignment. Maintenance philosophy and design are, nevertheless, interdependent. Most of the foregoing have some influence on the choice of test equipment. Skill requirements are influenced by the choice of LRA, by displays and by standardisation. Maintenance procedures are affected by the size of modules and the number of types of spares. The following chapter will examine the ways in which maintenance philosophy and design act together to influence down times.

6 Maintenance Philosophy and Down Time

Both active and passive repair times are influenced by factors other than equipment design. Consideration of maintenance procedures, personnel, and spares provisioning is known as Maintenance Philosophy and plays an important part in determining overall availability. The costs involved in these activities are considerable and it is therefore important to strike a balance between over and underemphasising each factor. They can be grouped under six headings:

Organisation of maintenance resources.
Tools and Test Equipment.
Personnel – selection, training and motivation.
Maintenance instructions and manuals.
Spares provisioning.
Logistics.

6.1 ORGANISATION OF MAINTENANCE RESOURCES

It is usual to divide the maintenance tasks into three groups in order firstly to concentrate the higher skills and more important test equipment in one place and secondly to provide optimum replacement times in the field. These groups, which are known by a variety of names, are as follows.

6.1.1 First Line Maintenance – Corrective Maintenance – Call – Field Maintenance

This will entail diagnosis only to the level of the LRA and repair is by LRA replacement. The technician either carries spare LRAs or has rapid access to them. Diagnosis may be aided by a portable intelligent terminal especially in the case of microprocessor based equipment. This group may involve two grades of technician, the first answering calls and the second small group of specialists who can provide backup in the more difficult cases.

6.1.2 Preventive Maintenance – Routine Maintenance

This will entail scheduled replacement of defined modules and some degree of cleaning and adjustment. Parametric checks to locate dormant faults and drift conditions may be included.

6.1.3 Second line Maintenance – Workshop – Overhaul Shop – Repair Depot

This is for the purpose of:

(a) Scheduled overhaul and refurbish of units returned from preventive maintenance.
(b) Unscheduled repair and/or overhaul of modules which have failed or become degraded.

A deeper diagnostic capability is implied and therefore the larger more complex test equipment will be found at the workshop together with full system information.

6.2 MAINTENANCE PROCEDURES

For any of the above groups of staff it has been shown that fast, effective and error-free maintenance is best achieved if a logical and formal procedure is followed on each occasion. A haphazard approach based on the subjective opinion of the maintenance technician, although occasionally resulting in spectacular short cuts, is unlikely to prove the better method in the long run. A formal procedure also ensures that calibration and essential checks are not omitted, that diagnosis always follows a logical sequence designed to prevent incorrect or incomplete fault detection, that correct test equipment is used for each task (damage is likely if incorrect test gear is used) and that dangerous practices are avoided. Correct maintenance procedure is only assured by accurate and complete manuals and thorough training. A maintenance procedure must consist of the following:

Making and interpreting test readings.
Isolating the cause of a fault.
Part (LRA) replacement.
Adjusting for optimum performance (where applicable).

The extent of the diagnosis is determined by the level of fault identification and hence by the Least Replaceable Assembly. A number of procedures are used:

(i)   Stimuli – response: where the response to changes of one or more parameter is observed and compared with the expected response.
(ii)   Parametric checks where parameters are observed at displays and test points and are compared with expected values.
(iii) Signal injection where a given pulse, or frequency, is applied to a particular

point in the system and the signal observed at various points, in order to detect where it is lost, or incorrectly processed.

(iv) Functional isolation wherein signals and parameters are checked at various points, in a sequence designed to eliminate the existence of faults before or after each point. In this way the location of the fault is narrowed down.

(v)   Robot test methods where automatic test equipment is used to fully 'flood' the unit with a simulated load, in order to allow the fault to be observed.

Having isolated the fault, a number of repair methods present themselves:

1. Direct replacement of the LRA.
2. Component replacement or rebuilding, using simple construction techniques.
3. Cannibalise from non-essential parts.

In practice direct replacement of the LRA is the usual solution due to the high cost of field repair and the need for short down times in order to achieve the required equipment availability.

Depending upon circumstances, and the location of a system, repair may be carried out either immediately a fault is signalled, or only at defined times, with redundancy being relied upon to maintain service between visits. In the former case, system reliability depends on the mean repair time and in the latter, upon the interval between visits and the amount of redundancy provided.

6.3 TOOLS AND TEST EQUIPMENT

The following are the main considerations when specifying tools and test equipment.

1. Simplicity: test gear should be easy to use and require no elaborate set-up procedure.
2. Standardisation: the minimum number of types of test gear reduces the training and skill requirements and minimises test equipment spares holdings. Standardisation should include types of displays and connections.
3. Reliability: test gear should be an order of magnitude more reliable than the system for which it is designed, since a test equipment failure can extend down time or even result in a system failure.
4. Maintainability: ease of repair and calibration will affect the non-availability of test gear. Ultimately it reduces the amount of duplicate equipment required.
5. Replacement: suppliers should be chosen bearing in mind the delivery time for replacements and for how many years they will be available.

There is a trade-off between the complexity of test equipment and the skill and training of maintenance personnel. This extends to built-in test equipment (BITE) which, although introducing some disadvantages, speeds and simplifies maintenance.

BITE forms an integral part of the system and requires no setting up procedure in order to initiate a test. Since it is part of the system, weight, volume and power consumption are improtant. A customer may specify these constraints in the system specification (e.g. power requirements of BITE not to exceed 2 per cent of mean power consumption). Simple BITE can be in the form of displays of various parameters. At the other end of the scale, it may consist of a programmed sequence of stimuli and tests, which culminate in a 'print out' of diagnosis and repair instructions. There is no simple formula, however, for determining the optimum combination of equipment complexity and human skill. The whole situation, with the variables mentioned, has to be considered and a trade-off technique found which takes account of the design parameters together with the maintenance philosophy.

There is also the possibility of Automatic Test Equipment (ATE) being used for field maintenance. In this case the test equipment is quite separate from the system and is capable of monitoring several parameters simultaneously and on a repetitive basis. Control is generally by software and the maintenance task is simplified. ATE is discussed in some depth in chapter 9.

When choosing simple portable test gear, there is a choice of commercially available general purpose equipment, as against specially designed equipment. Cost and ease of replacement favour the general purpose equipment whereas special purpose equipment can be made simpler to use and more directly compatible with test points.

In general, the choice between the various test equipment options involves a trade-off of complexity, weight, cost, skill levels, time scales and design, all of which involve cost, with the advantages of faster and simpler maintenance.

6.4 PERSONNEL CONSIDERATIONS

Four manpower considerations influence the maintainability of an equipment:

Training given.
Skill level employed.
Motivation.
Quantity and distribution of personnel.

More complex designs involve a wider range of maintenance and hence more training is required. Proficiency in carrying out corrective maintenance is achieved by a combination of knowledge and diagnostic skill. Whereas knowledge can be aquired by direct teaching methods, skill can be only be gained from experience, either in a simulated or real environment. Training must, therefore, include experience of practical fault finding on actual equipment. Sufficient theory, in order to understand the reasons for certain actions and to permit logical reasoning, is required, but an excess of theoretical teaching is both unnecessary and confusing. A balance must be achieved between the confusion of too much theory and the motivating interest created by such knowledge.

A problem with very high reliability equipment is that some failure modes occur so infrequently that the technicians have little or no field experience of their diagnosis and repair. Refresher training with simulated faults will be essential to ensure effective maintenance, should it be required. Training maintenance staff in a variety of skills (e.g. electronic as well as electro-mechanical work) provides a flexible work force and reduces the probability of a technician being unable to deal with a particular failure unaided. Less time is wasted during a repair and transport costs are also reduced.

Training of customer maintenance staff is often given by the contractor, in which case an objective test of staff suitability may be required. Well-structured training which provides flexibility and proficiency, improves motivation since confidence, and the ability to perform a number of tasks, brings job satisfaction in demonstrating both speed and accuracy. In order to achieve a given performance, specified training and a stated level of ability are assumed. Skill levels must be described in objective terms of knowledge, dexterity, memory, visual acuity, physical strength, inductive reasoning and so on.

Manpower scheduling requires a knowledge of the equipment failure rates. Different failure types require different repair times and have different failure rates.

The MTTR may be reduced by increasing the effort from one to two technicians but any further increase in manpower may not significantly reduce the repair time.

Personnel policies are usually under the control of the customer and, therefore, close liaison between contractor and customer is essential before design features relating to maintenance skills can be finalised. In other words the design specification must reflect the personnel aspects of the maintenance philosophy.

6.5 MAINTENANCE INSTRUCTIONS

The most vital repair tool is the maintenance manual. It has to be accurate and complete and the information easily located otherwise the manual is likely to be discarded for being more a hindrance than a help. Its main use is as an aid to diagnosis for which it should outline a logical sequence of tests necessary to identify, by a process of elimination, the cause of a malfunction. Wherever possible procedures should be self-checking by ensuring that the omission of a test or alignment will be detected by a later test. Functional block diagrams assist fault finding by providing a logical illustration of the checking sequence. Illustrations of correct and incorrect waveforms assist in fault cause recognition.

Other features of the manual should include an emphasis on safety hazards and precautions, details of preventive maintenance, spares requirements and instructions for failure reporting. Failure reporting causes a traditional conflict between those who seek to interpret fault data and those who are too busy to provide it.

Efforts should be made to anticipate maintenance situations where damage could result from the maintenance action. These hazards should be clearly, identified. Searching for dormant faults and degradation conditions is time consuming and therefore expensive. Wherever possible these checks should be carried out when access has already been gained for some other essential maintenance task.

A manual must be clear and easy to follow if the technician is to be convinced of its value and should contain no unnecessary theory. Chapter 10 elaborates further on the layout and presentation of handbooks.

6.6 SPARES PROVISIONING

The number of identical units for which a particular spare is to be kept, together with a knowledge of the failure rate of that item, will yield the likely required number and hence cost, of that particular spare. Since failure rate is a statistical quantity the cost of spares obtained in this way, will not be an exact forecast. Allowance must also be made for spares damaged in transport, or by handling, for those which fail in store, and for those which may be used in error, where there is no corresponding part failure.

The number of spares predicted is related to the risk of a stockout. Assuming a random distribution of failures, the number of spares can be forecast as shown in figure 6.1.

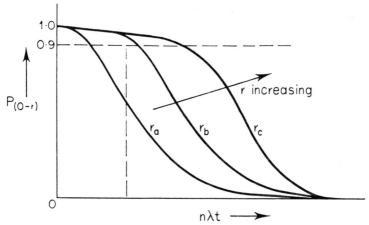

Figure 6.1   Set of curves for spares provisioning.
λ = failure rate
n = number of parts of that type which may have to be replaced
r = number of spares of that part carried
$P_{(0-r)}$ = probability of $0 - r$ failures = probability of stock not being exhausted

It assumes that the failures are at random and that therefore a set of Poisson curves can be used. The figure shows how P_{0-r} varies with $n\lambda t$ for different values of r. It shows how a specific value of $n\lambda t$ with a 10 per cent risk of stockout is associated with a spares holding of r_b.

Another situation occurs where the stock of spares is continually replenished by repaired failed items. Under these circumstances the MTTR of the failed item is relevant. In this case the repair time refers to the second-line repair and not to the MTTR at the system level. It is assumed that system repair is effected by replacement of a failed unit which is then repaired 'off-line' and returned to the stock of spares. This is illustrated in figure 6.2.

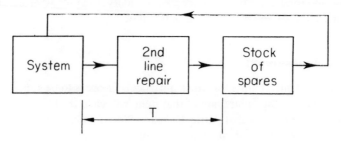

Figure 6.2   Spares replacement from second-line repair

The question arises as to whether spares that have been repaired should be returned to a central stock or retain their identity for return to the parent system. Returning a part to its original position is costly and requires a procedure so that initial replacement is only temporary. This may be necessary where servicing is carried out on equipment belonging to different customers — indeed some countries impose a legal requirement to this end. Another reason for retaining a separate identity for each unit occurs due to wearout, when it is necessary to know the expired life of each item.

Stock control is necessary when holding spares and inputs are therefore required from:

Preventive and corrective maintenance in the field.
Second line maintenance.
Warranty items supplied.

The main considerations of spares provisioning are:

1. Failure rate — determines quantity and perhaps location of spares.
2. Acceptable probability of stockout — fixes spares level.
3. Turn round of second line repair — effects spares holding.
4. Cost of each spare — effects spares level and hence item 2.

5. Standardisation and LRA — effects number of different spares to be held.
6. Lead time on ordering — effectively part of second line repair time.

6.7 LOGISTICS

Logistics is concerned with the time and resources involved in transporting men, spares and equipment into the field. The main consideration is the degree of centralisation of these resources.

Centralise	*Decentralise*
Specialised test equipment.	Small tools and standard items.
Low utilisation skills and testgear.	Where small MTTR is vital.
Second line repair.	Fragile test gear.
Infrequent (high reliability) spares.	Frequent (low reliability) spares.

A combination will be found where a minimum of on-site facilities, which ensure repair within the specified MTTR, are provided. The remainder of the spares backup and low utilisation testgear can then be centralised. If Availability is to be kept high by means of a low MTTR then spares depots may have to be established at a sufficient number of points to permit access to spares within a specified time.

6.8 THE USER AND THE DESIGNER

The considerations discussed in this chapter are very much the user's concern. It is necessary, however, to decide upon them at the design stage since they influence, and are influenced by, the engineering of the product. The following table shows a few of the relationships between maintenance philosophy and design.

Skill level of maintenance technician	Amount of built-in test equipment required
	Level of LRA replacement in the field
Tools and test equipment	LRA fixings, connections and access
	Test points and equipment standardisation
Maintenance procedure	Ergonomics and environment
	Built-in test equipment diagnostics
	Displays
	Interchangeability

The importance of user involvement at the very earliest stages of design cannot be overemphasised. Maintainability objectives cannot be satisfied merely by placing requirements on the designer and neither can they be considered without recognising that there is a strong link between repair time and cost. The maintenance philosophy has therefore to be agreed whilst the design specification is being prepared.

7 Analysis of Failure Mode and Stress

7.1 STRESS AND FAILURE

The probability of a device failing at any instant is very sensitive to the stress applied to it. Stresses, which can be classified as environmental or self generated, include:

Temperature
Shock
Vibration   } Environmental
Humidity
Foreign bodies

Power dissipation
Applied voltage   } Self-generated
Self-vibration and force

The overall sum of these stresses is often pictured as constantly varying, having peaks and troughs, and superimposed on a distribution of strength levels for a group of devices. A failure is assumed to be the result of stress exceeding strength. The average strength of the group of devices will increase during the early failures period due to the elimination, from the population, of the weaker items. During wearout strength declines as a result of physical and chemical processes. An overall change of the average stress will cause more of the peaks to exceed the strength values and more failures will result. Figure 7.1 illustrates this concept showing a range of strength throughout the bathtub together with a superimposed strain.

7.2 FAILURE MODE ANALYSIS

There are two techniques to be found under this heading:

Failure Mode Effect and Criticality Analysis	*and*	*Fault Tree Analysis*
Sometimes called Stress and Failure analysis. Analyses component failure rates and stresses and compute system reliability (FMECA).		Defines all possible system faults and traces causes and probabilities. A top down approach.

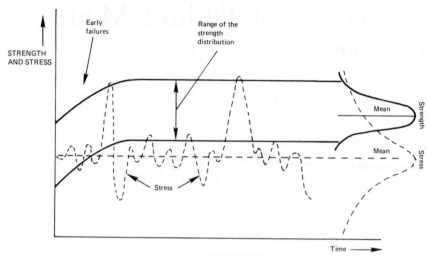

Figure 7.1   Strength and stress

7.2.1 FMECA

Failure Mode Effect and Criticality Analysis is a systematic process whereby faults at the part/component level are identified and, using recorded failure rates at the appropriate stress levels, their effect at the system level is determined. Each part is considered, in turn, as having failed in each possible mode. The effect of each of these imaginary failures at various system levels is noted and a failure rate assigned to it from available data. Each system level failure will have resulted from various possible component failures and these can be grouped together for the purpose of calculating the system failure rate. By ranking or weighting system failure modes in some order of importance it is possible to assign priorities for design attention at the component and circuit levels. Figure 7.2 shows a typical stress and failure analysis worksheet for recording the details and failure rate of each component. Failure rates are obtained from sets of data such as US MIL Handbook 217C. Appendix 3 gives some typical failure rates for general guidance. They have been compiled from a number of sources and do not represent any one particular data source. Their purpose is to provide the reader with a general picture of the types of component failure rate and their variation with temperature, stress, environment and quality level.

7.2.2 Fault Tree Analysis

This involves identifying system level faults and then constructing a logic diagram showing all the possible combinations of failures and conditions which lead to each. Failure mode probabilities are then computed from basic fault data. This method is considered most applicable during design and pre-production stages of a product. The result is a logic flow diagram as shown in Fig. 7.3 and a

UNIT _P/5_ SYSTEM _4830_ DATE _1-1-80_ ENGINEER _RPR_ SHEET _1_ of _5_

AMBIENT TEMPERATURE _30_ °C

PART DESCRIPTION	Actual	Rated	Ratio	Failure Mode	Effect on Circuit	Effect on System	Criticality	Basic λ at Ambient Temp. and stress	QA and Envirn't	Special factor	Failure rate 10^{-9}	Comment
R.23 Metal Oxide	5w/w	1w	5	S/C	No op	No Alarm	1*	3	1	1	3	

TOTAL FAILURE RATE
_____

*Criticality code eg Safety hazard

Figure 7.2   Stress and failure analysis worksheet

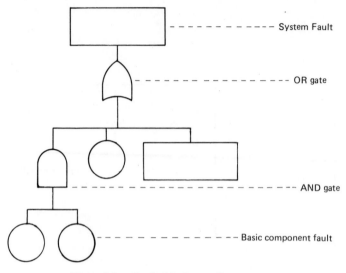

Figure 7.3   Typical fault tree diagram

tabulated list of fault conditions, with probabilities, showing the areas for corrective design or part selection required.

7.3 FAILURE MECHANISMS

7.3.1 Types of Failure Mechanism

The majority of failures are attributable to one of the following physical or chemical phenomena.

Alloy Formation   Formation of alloys between gold, aluminium and silicon causes what is known as 'purple plague' and 'black plague' in silicon devices.

Biological Effects   Moulds and insects can cause failures. Tropical environments are particularly prone and electronic devices and wiring can be effected.

Chemical and Electrolytic Changes   Electrolytic corrosion can occur wherever a potential difference together with an ionisable film is present. The electrolytic effect causes interaction between the salt ions and the metallic surfaces which act as electrodes. Salt laden atmospheres cause corrosion of contacts and connectors. Chemical and physical changes to electrolytes and lubricants both lead to degradation failures.

Contamination   Dirt, particularly carbon or ferrous particles, causes electrical failure. The former deposited on insulation between conductors leads to breakdown and the latter to insulation breakdown and direct short circuits. Non-conducting material such as ash and fibrous waste can cause open circuit failure in contacts.

Depolymerisation   This is a degrading of insulation resistance caused by a type of liquefaction in synthetic materials.

Electrical Contact Failures   Failures of switch and relay contacts occur due to weak springs, contact arcing, spark erosion and plating wear. In addition failures due to contamination, as mentioned above, are possible. Printed board connectors will fail due to loss of contact pressure, mechanical wear from repeated insertions and contamination.

Evaporation   Filament devices age due to evaporation of the filament molecules.

Fatigue   A physical/crystalline change in metals leading to spring failure, fracture of structural members, etc.

Film Deposition   All plugs, sockets, connectors and switches with non-precious metal surfaces are likely to form an oxide film which is a poor conductor. This film therefore leads to high resistance failures unless a self-cleaning, wiping action is used.

Friction   Friction is one of the most common causes of failure in motors, switches, gears, belts, styli, etc.

Ionisation of Gases   At normal atmospheric pressure a.c. voltages of approximately 300 V across gas bubbles in dielectrics give rise to ionisation which causes both electrical noise and ultimate breakdown. This reduces to 200 V at low pressure.

Ion Migration   If two silver surfaces are separated by a moisture covered insulating material then, providing an ionisable salt is present as is usually the case, ion migration causes a silver 'tree' across the insulator.

Magnetic Degradation   Modern magnetic materials are quite stable, however degraded magnetic properties do occur as a result of mechanical vibration or strong a.c. electric fields.

Mechanical Stresses   Bump and vibration stresses affect switches, insulators, fuse mountings, component lugs, printed board tracks, etc.

Metallic Effects   Metallic particles are a common cause of failure as mentioned above. Tin and Cadmium can grow 'whiskers' leading to noise and low resistance failures.

Moisture Gain or Loss   Moisture can enter equipment through pin holes by moisture vapour diffusion. This is accelerated by conditions of temperature cycling under high humidity. Loss of moisture by diffusion through seals in electrolytic capacitors causes reduced capacitance.

Molecular Migration   Many liquids can diffuse through insulating plastics.

Stress Relaxation   Cold flow ('creep') occurs in metallic parts and various dielectrics under mechanical stress. This leads to mechanical failure. This is not the same as fatigue which is caused by repeated movement (deformation) of a material.

Temperature Cycling   This can be the cause of stress fluctuations leading to fatigue or to moisture build up.

7.3.2 Failures in Semiconductor Components

The majority of semiconductor device failures are attributable to the wafer fabrication process. The tendency to create chips with ever decreasing cross sectional areas increases the probability that impurities, localised heating, flaws, etc., will lead to failure by deterioration, probably of the Arrenhius type (see section 4.2.2). Table 7.1 shows a typical division of failure modes.

7.3.3 Discrete Components

The most likely causes of failure in resistors and capacitors are shown in tables 7.2 and 7.3.

Table 7.1

	Specific			In general
	Linear	*TTL*	*CMOS*	
	%	%	%	%
Metallisation	18	50	25	
Diffusion	1	1	9	55
Oxide	1	4	16	
Bond–die	10	10	—	
Bond–wire	9	15	15	25
Packaging/hermeticity	5	14	10	
Surface contamination	55	5	25	20
Cracked die	1	1	—	

Table 7.2

Resistor type	Short	Open	Drift
Film	Insulation breakdown due to humidity Protuberances of adjacent spirals	Mechanical breakdown of spiral due to r.f. Thin spiral	–
Wire would	Over voltage	Mechanical breakdown due to r.f. Failure of winding termination	
Composition			r.f. produces capacitance or dielectric loss
Variable (wire and composition)		Wiper arm wear Excess current over a small segment due to selecting low value	Noise Mechanical movement

Table 7.3

Capacitor type	Short	Open	Drift
Mica	Water absorbtion Silver ion migration	Mechanical vibration	
Electrolytic solid tantalum	Solder balls caused by external heat from soldering	Internal connection Failures due to shock or vibration	
Electrolytic non-solid tantalum	Electrolyte leakage due to temperature cycling	External welds	
Electrolytic aluminium oxide		Lead dissolved in electrolyte	Low capacitance due to aluminium oxide combining with electrolyte
Paper	Moisture Rupture	Poor internal connections	
Plastic	Internal solder flow Instantaneous breakdown in plastic causing s/c	Poor internal connections	
Ceramic	Silver ion migration	Mechanical stress Heat rupture internal	
Air (variable)	Loose plates Foreign bodies	Ruptured internal connections	

Short circuit failure is rare in resistors. For composition resistors, fixed and variable, the division tends to be 90 per cent degradation failures and 10 per cent open circuit. For film and wirewound resistors the majority of failures are of the open circuit type.

7.4 ENVIRONMENTAL STRESSES AND FAILURE RATE

Although empirical relationships have been established relating certain device failure rates to specific stresses, such as voltage and temperature, no precise formula exists which links specific environments to failure rates. The permutation of different values of environmental factors such as are listed in section 8.2. is immense. General adjustment (multiplying) factors have been evolved and these are used to scale up basic failure rates to particular environmental conditions. The best known of these are the environmental factors listed in US MIL Handbooks 217B and C. The condition, for which the multiplier is unity, is a fixed ground installation with reasonable atmospheric conditions and average maintenance. More benign or more severe environments are ascribed factors less or greater than one, accordingly. These factors, which are developed from field data, are specific to each of various classes of component. Different part types are affected to different degrees by each type of environment and the range of multiplying factors covers 0.2 to 120 times the basic failure rates. Table 7.4 shows various factors for a range of defined environments for monolithic devices.

Table 7.4

Environmental factors for monolithic microelectronics US MIL 217C

Environment	Conditions	Multiplier
Ground benign (G_B)	Nearly zero stress with optimum engineering and operation and maintenance	0.2
Space flight (S_F)	Earth orbital. Approaches Ground Benign conditions without access for maintenance. Vehicle neither under powered flight nor in re-entry.	0.2
Ground fixed (G_F)	Conditions less than ideal. Installation in permanent racks with adequate cooling air. Maintenance by military personnel and installation in unheated buildings	1.0

Environment	Conditions	Multiplier
Ground mobile (G_M) (and portable)	Conditions more severe than those for fixed ground; mostly for vibration and shock. Cooling air supply may also be more limited, and maintenance less uniform	4.0
Naval, sheltered (N_S)	Surface ship conditions similar to fixed ground, but subject to occasional high shock and vibration	4.0
Naval, unsheltered (N_U)	Nominal surface shipborne conditions, but with repetetive high levels of shock and vibration	5.0
Airborne inhabited (A_{IT})	Typical cockpit conditions without environmental extremes of pressure, temperature, shock and vibration. NB (A_{IF}) for fighter	2.8
Airborne uninhabited (A_{UT})	Bomb-bay, tail, or wing installations where extreme pressure, temperature and vibration cycling may be aggravated by contamination from oil, hydraulic fluid and engine exhaust. NB (A_{UF}) for fighter	4.2
Missile launch (M_L)	Severe conditions of noise, vibration, and other environments related to missile, launch and space vehicle boost into orbit, vehicle re-entry and landing by parachute. Conditions may also apply to installation near main rocket engines during launch operations	10.0

7.5 FAILURE RATE DATA

Many independent collections of failure rate data now exist. These have been compiled, from field and laboratory experience by defence departments, telecommunications authorities and private companies. The most well known published data is US MIL Handbook 217C which is both detailed and thorough. Failure rate data is usually, unless otherwise specified, taken to refer to random failures and hence constant failure rate applies. It is important to read, carefully, any covering notes since, for a given temperature and environment, a component may exhibit a wide range of failure rates due to:

(i)   Component source — the degree of screening, QA etc.
(ii) Circuit tolerancing — the degree of design effort affects the proportion of failures attributable to parametric drift.
(iii) Reliability growth — the amount of field experience fed back affects reliability and hence the failure rate data.
(iv) Maintenance philosophy — trial and error replacement, as a means of diagnosis, artificially inflates failure rate data.

Failure rate values can span one and two orders of magnitude as a result of different combinations of these factors. Prediction calculations are explained in chapter 15 but it will be emphasised that the relevance of failure rate data is more important than refinements in the statistics of the calculation.

The failure rates are usually tabulated, for a given component type, against ambient temperature and the ratio of applied to rated stress (power or voltage). Appendix 3 contains generic failure rate data from US MIL Handbook 217C and shows an average failure rate for each component tabulated against environment. The main body of the Handbook gives the data against temperature and applied stress. Having selected a failure rate a multiplying factor is applied to take account of the quality level of the component.

8 Design and Qualification Testing

There are three categories of testing:

Design Testing – Laboratory and prototype tests aimed at proving that a design
will meet the specification. Initially bread board functional tests aimed at
proving the design. This will extend to preproduction models which undergo
environmental and reliability tests and may overlap with:

Qualification Testing – Total proving cycle using production models over the full
range of the environmental and functional specification. This involves
extensive marginal tests, climatic and shock tests, reliability and maintainability tests and the accumulation of some field data. It must not be confused
with development or production testing. There is also:

Production Testing and Commissioning – Verification of conformance by testing
modules and complete equipment. Some reliability proving and burn in may
be involved. Generally failures will be attributable to component procurement,
production methods, etc. Design related queries will arise but should diminish
in quantity as production continues.

Acceptance testing implies a formal demonstration and may apply to qualification
or to production tests depending upon the circumstances. In the former case a
contract development may lead to a formal demonstration of design conformance
and in the latter, equipment already in manufacture may undergo demonstration
tests for reasons of quality audit or customer inspection.

8.2 ENVIRONMENTAL TESTING

This involves proving that an equipment functions to specification (for a
sustained period) and is not degraded or damaged by defined extremes of its
environment. This can cover a wide range of parameters and it is important to
agree a specification which is realistic. It is tempting, when in doubt, to widen
the limits of temperature, humidity and shock in order to be sure of covering
the likely range which the equipment will experience. The resulting cost of
overdesign, even for a few degrees of temperature, may be totally unjustified.

The possibilities are numerous and include:

ELECTRICAL
>   Electric fields.
>   Magnetic fields.

CLIMATIC
>   Temperature extremes ⎫
>   Temperature cycling   ⎬ internal and external may be specified.
>   ⎭
>   Humidity extremes.
>   Temperature cycling at high humidity.
>   Thermal shock — rapid change of temperature.
>   Wind — both physical force and cooling effect.
>   Wind and precipitation.
>   Direct sunlight.
>   Atmospheric pressure extremes.

MECHANICAL
>   Vibration at given frequency — a resonant search is often carried out.
>   Vibration at simultaneous random frequencies — used because resonances at
>   different frequencies can occur simultaneously.
>   Mechanical shock — bump.
>   Acceleration.

CHEMICAL
>   Corrosive atmosphere — covers acids, alkalies, salt, greases, etc.
>   Foreign bodies — ferrous, carbon, silicate, general dust, etc.
>   Biological — defined growth or insect infestation.
>   Reactive gases.

8.3 MARGINAL TESTING

This involves proving the various system functions at the extreme limits of the
electrical and mechanical parameters. This includes:

ELECTRICAL
>   Mains supply voltage.
>   Mains supply frequency.
>   Insulation limits.
>   Earth testing.
>   High voltage interference — radiated. Typical test apparatus consists of a
>   spark plug, induction coil and break contact.
>   Mains borne interference.
>   Line error rate — refers to the incidence of binary bits being incorrectly
>   transmitted in a digital system. Usually expressed as 1 in 10^{-n} bits.
>   Line noise tests — analogue circuits.

Functional load tests — loading a system with artificial traffic to simulate full utilisation (e.g. call traffic simulation in a telephone exchange).

Input/output signal limits — limits of frequency and power.

Output load limits — sustained voltage at maximum load current and testing that current does not increase even if load is increased as far as a short circuit.

MECHANICAL

Dimensional limits — maximum and minimum limits as per drawing.

Pressure limits — covers hydraulic and pneumatic systems.

Load — compressive and tensile forces and torque.

Acceleration — linear and rotary.

8.4 HIGH RELIABILITY TESTING

The major problem in verifying high reliability, as will be emphasised in chapter 12, is the difficulty of accumulating sufficient data, even with no failures, to demonstrate statistically the value required. If an MTBF of say 10^6 h is to be verified, and 500 equipments are available for test, then 2000 elapsed hours of testing (3 months of continuous test) are required to accumulate sufficient time for even the minimum test which involves no failures. In this way the MTBF is demonstrated with 63 per cent confidence. Nearly $2\frac{1}{2}$ times the amount of testing is required to raise this to 90 per cent.

The usual response to this problem is to accelerate the failure mechanisms by increasing the stress levels. This involves the assumption that relationships between failure rate and stress levels hold good over the range in question. Interpolation between points in a known range presents little problem whereas extrapolation beyond a known relationship is of dubious value. Experimental data can be used to derive the constants found in the equations shown in section 4.2.2. In order to establish if the Arrhenius relationship applies a plot of $\log_e$ failure rate against the reciprocal of temperature is made. A straight line indicates that it holds for the temperature range in question. In some cases parameters such as temperature and power are not independent as in transistors where the junction temperature is a function of both. Accelerated testing gives a high confidence that the failure rate at normal stress levels is, at least, less than that observed at the elevated stresses.

Where MTBF is expressed in cycles or operations as with relays, pistons, rods and cams the test may be accelerated without a change in the physics of the failure mechanism. For example 100 contactors can be operated to accumulate 3×10^8 operations in one month although, in normal use, it might well take several years to accumulate the same amount of data.

8.5 RELIABILITY GROWTH

This is defined as the improvement in field reliability, during use, due to field data feedback resulting in modifications. This improvement depends on reliability management ensuring that field data leads to design modifications. It follows that the majority of failure types thus eliminated are design related. Early field reliability is usually less than the level predicted since assessed reliability only acknowledges random component failures. Growth, then, is the process of eliminating design related failures. It must not be confused with the decreasing failure rate observed in the bathtub curve which applies to any equipment irrespective of the degree of reliability growth. Figure 8.1 illustrates the difference with two bathtub curves for the same equipment. Both exhibit an early (decreasing failure rate) failures portion whereas the later product, subject to some reliability growth, shows a higher reliability in the random failures area of the curve.

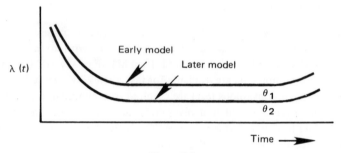

Figure 8.1

There is clearly a benefit in relating θ_1 and θ_2 to the total accumulated field time since this would enable predictions of growth for a given amount of field experience to be carried out. Conversely the likely elapsed field time required to achieve a given improvement could be calculated. The rate of growth is a function of the amount of development effort applied to field failure data. The growth factor is not usually known at the beginning and, in practice, any prediction is refined as data are accumulated.

The best known growth/time relationship is that described by J. T. Duane, in 1962, for electronic and electromechanical equipment. It assumes the empirical relationship whereby the improvement in MTBF is proportional to T^α where T is the total equipment time and α is a growth factor, typically between 0.1 and 0.65. Figure 8.2 shows a Duane plot of cumulative MTBF against cumulative time on log axes.

r shows the range of cumulative time required to achieve a specified MTBF for factors in the range 0.2–0.5.

Changes resulting from a growth programme are likely to affect:

Design – components, parts, tolerances, loading
Production methods and processes.
Testing and installation methods.

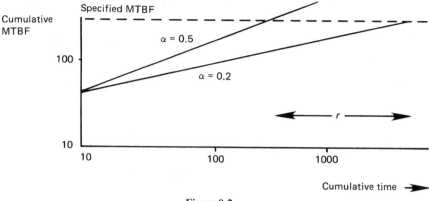

Figure 8.2

Packaging.
Operational use of the equipment.
Preventive maintenance routines.

8.6 TESTING FOR PACKAGING AND TRANSPORT

There is little virtue in investing large sums in design and manufacture if
inherently reliable products are to be damaged by inadequate packaging and
handling. The packaging needs to match the characteristics and weaknesses of the
contents with the hazards it is likely to meet. The major causes of defects during
packaging, storage and transport are:

(a) Inadequate or unsuitable packaging materials for the transport involved.
     Transport, climatic and vibration conditions not foreseen.
     Storage conditions and handling not foreseen.
     – requires consideration of waterproofing, hoops, bands, lagging, hermetic
       seals, desiccant, ventilation holes, etc.
(b) Inadequate marking – see British Standard 2770 – Pictorial Handling
     Instructions.
(c) Failure to treat for prevention of corrosion.
     – various cleaning methods for the removal of oil, rust and miscellaneous
       contamination followed by preventive treatments and coatings.
(d) Degradation of packaging materials due to method of storage prior to use.
(e) Inadequate adjustments or padding prior to packaging.
     Lack of handling care during transport.
     – requires adequate work instructions, packing lists, training, etc.

Choosing the most appropriate packaging involves considerations of cost,
availability and size for which reason a compromise is usually sought. Crates,

rigid and collapsible boxes, cartons, wallets, tri-wall wrapping, chipboard cases, sealed wrapping, fabricated and moulded spacers, corner blocks and cushions, bubble wrapping, etc., are a few of the many alternatives available to meet any particular packaging specification.

Environmental testing involving vibration and shock tests together with climatic tests is necessary to qualify a packaging arrangement. This work is undertaken by a number of test houses and may save large sums if it ultimately prevents damaged goods being received since the cost of defects rises tenfold, and more, once equipment has left the factory. As well as specified environmental tests the product should be transported over a range of typical journeys and then retested to assess the effectiveness of the proposed pack.

8.7 MULTIPARAMETER TESTING

More often than not the number of separate, but not independent, variables involved in a test makes it impossible for the effect of each to be individually assessed. To hold, in turn, all but one parameter constant and record its effect and then to analyse and relate all the parametric results would be very expensive in terms of test and analysis time. In any case this has the drawback of restricting the field of data. Imagine that, in a three variable situation, the limits are represented by the corners of a cube as in figure 8.3 then each test would be confined to a straight line through the cube.

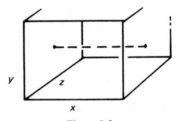

Figure 8.3

One effective approach involves making measurements of the system performance at various points, including the limits, of the cube. For example in a facsimile transmission system the three variables might be the line error rate, line bandwidth and degree of data compression. For each combination the system parameters would be character error rate on received copy and transmission time. Analysis of the cube would reveal the best combination of results and system parameters for a cost effective solution.

8.8 TEST HOUSES

Many of the types of test described in this chapter are carried out, commercially, by test houses. It is necessary to find a house whose expertise and equipment match the type of testing required. One must enquire if the range of facilities can cope with the sizes and weights of equipment involved as well as with the combinations of environmental factors. The UK Department of Prices and Consumer Protection has issued a Register of Test Houses (1976) which contains an alphabetical list of many test houses together with a cross index of the types of product testing and facilities offered.

9 Quality Assurance and Automatic Test Equipment

9.1 FUNCTIONS OF QUALITY ASSURANCE

Quality is simply defined as conformance to specification. It is not, in the engineering context, a measure of excellence. The simple and inexpensive item which conforms to specification is therefore of 'higher' quality than an elaborate and expensive item which does not. The purpose of Quality Assurance is to set up and operate a set of controls whereby the appropriate activities in design, procurement and manufacturing, test, installation and maintenance are carried out to ensure that products meet the specification at MINIMUM cost (see chapter 2). These activities fall into four areas.

9.1.1 Quality and Reliability Planning

(a) Test and Inspection Procedures
   Should be written at the design stage if possible since this activity can contribute to testability and hence cost reduction.
   Test problems referred to the test planning engineers can be fed back to the design resulting in reliability growth. Particularly relevant for adjustment and tolerance queries.
   Automatic test equipment programming has strong links with design, see section 9.3.
(b) Test Equipment Design
   Challenges testability/maintainability especially if dealt with at the design stage.
(c) Calibration
   Requires a study of the design in order to establish measurement accuracy requirements.
   Requires administration for recall and a method of varying the periodicity of recall according to the degree of drift measured at each calibration.
(d) Standards
   Process and workmanship standards covering design, manufacture, packaging, etc., e.g. design of printed boards; inspection of PCBs; hand solder joints; wrapping; crimping; packaging of component parts.

(e) Procedures

The Quality Manual is the top level document outlining policy, company and QA organisation and describes general procedures. It should be concise and not contain detail procedures.

Detailed quality procedures e.g. concessions, material review, change notes, vendor rating, calibration, audit.

(f) Audit

System audits cover a review of all activities in the company to see if they conform to the procedures/manuals. This is an important feature of formal QA schemes such as the UK Ministry of Defence Standards 0521 and 0524 which require comprehensive quality procedures and regular reviews.

Process and product audits are random checks by staff who are independent of test and inspection personnel.

Safety audits of processes and of the product are assuming greater emphasis with the advent of the Health and Safety Act and the emphasis on product liability.

(g) Training

Off the job training of foremen and operators should contain quality awareness information as, for example, practical information on quality costs. Practical techniques such as Pareto analysis of defects, statistical quality control charts and other sampling methods.

Training and knowledge are not likely to have any useful outcome unless the opportunity to put them into practice is available. Training material should therefore be relevant to the job.

(h) Quality Costs

Essentially an accounting responsibility although the impetus needs to come from quality management.

9.1.2 Quality Control

(a) Test and Inspection

This is the measurement or 'appraisal' aspect of QA. It involves visual and mechanical measurement to drawings and electrical and functional tests to a specification.

Inspection involves First Off; Patrol; Batch; and 100 per cent procedures according to the quantity and type of production involved.

Testers should work to a full test specification. Rework should be minimal with module replacement being used to rectify equipment. To hold a product pending minor rework ties up capital and reduces output.

Automatic test equipment is usually highly cost effective.

Thorough defect reporting is necessary and a standard printed format is preferred.

The following list of defect types is typical:

Production/Assembly defects	Wrong component type
	Wrong component dimension
	Component wrongly mounted
	Wiring routing or dressing fault
	Incorrect termination of wire
	Wire wrap defect
	Soldering defect
	Wire crimp defect
	Damage to component or wire
	Incorrect adjustment
	Incorrect issue of printed board or module
Design related defect	Adjustment too critical to achieve spec.
	Drift during test
	Parts to drawing but no fit
Process	Impregnation defects
	Printed board track or soldering fault
	Dimensions of numerically controlled machined parts
Material defects	Type, gauge, finish, corrosion, age related distortion, etc.

A printed test sheet will be issued with each item to be tested and will contain such of the above classifications as are applicable to the product together with any others which experience indicates are relevant.

(b) Statistical Methods

Where the statistical distribution of some process variable or attribute is known then the control chart technique is a valuable tool in controlling the output quality. Consider a process manufacturing lamps whose light output varies according to the normal distribution. The light output of a device chosen, at random, at regular intervals can be plotted on a chart as shown in figure 9.1.

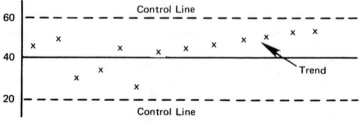

Figure 9.1   Control chart

If the horizontal axis represents the mean value of the output and larger and smaller values are represented by points above and below this line then the sample values may be plotted against time as shown. Two advantages are obtained. Firstly any persistent trend in the reading will be evident before the process limits are exceeded. Secondly the chart may have control lines indicating the probability of reaching a given value above or below the average assuming that the process is still under control. These statistical control lines are set such that the probability of a sample occurring that far from the average is a particular value, typically 1/40 (known as the warning line) or 1/1000 (known as the action line). It is argued that if a sample reaches one of these lines it is more likely that the process has gone out of control than that the sample value has occurred due to chance. Appropriate actions as defined by the two lines are then taken.

Sampling methods involve single and double samples and sequential sampling plans. All involve making statistical inferences about a batch of items on the basis of samples. The trade-off is between the cost of inspection and the degree of uncertainty in the inferred measurement.

9.1.3 Quality Engineering

(a) Reliability Engineering

This involves design reviews, reliability test planning, reliability prediction, field failure analysis and all the other activities described in this book. Chapter 17 deals with reliability and maintainability activities and management.

(b) Product Qualification

As discussed in chapter 8, a variety of tests to verify that a design conforms to the specification.

(c) Manufacturing Controls

Departures from specification/drawing may sometimes be permitted during manufacture. These require a concession procedure whereby a quality engineer investigates each application and places action requirements on engineering/production/industrial engineering, etc., as applicable.

The implementation of engineering changes in production and in the field must be controlled by a change coordinator whose responsibility it is to check stock, work in progress, installation and field items. Test and inspection procedures may need to be updated as a result of changes.

It is usual to have a Material Review Committee, chaired by a Quality Engineer, for the purpose of investigating non-conforming parts and assemblies. The purpose of the MRC is to:

(i) Sentence items.

(ii) Establish the cause of non-conformance.

(iii) Lay down measures to prevent a recurrence.

(iv) Investigate possible recovery of the item by rework.

(v) Record data of scrap and rework costs.

(d) Vendor Controls

Vendor assessment is carried out to establish the capacity of a supplier or subcontractor to meet the specification. This usually involves an audit of his procedures and controls by QA and purchasing staff. Suppliers are then given a grading which determines the extent of incoming inspection applied to their deliveries. A high rating is desirable since little or no incoming inspection means faster release of Goods Received Notes leading to prompt payment and, from the buyer's point of view, inspection costs are reduced. Vendor rating involves analysis of incoming inspection records in order to continuously update the supplier's grading. This may involve an assessment not only of quality but of price, delivery, packaging and overall cooperation (i.e. acceptance of small orders, engineering changes, inadequate drawings, etc.).

(e) Defect Analysis and Feedback

Defects should be recorded at incoming inspection, production, test and in the field. This effort is wasted unless the data is analysed and put to use for defect prevention. Statistics should be prepared on a weekly basis and regularly examined for trends. A Pareto analysis of defect costs is a powerful method of focusing attention on the major problems. A list of defects is prepared similar to the example of a test sheet in section 9.1.2. The frequency of each defect is recorded together with an estimate of the cost. The frequency of each defect type is then multiplied by the cost to provide a total cost penalty for each type. The defect types are then listed in order of total cost. In this way the most expensive group of defects and not necessarily the most frequent defect heads the list. The following simple example shows a Pareto analysis of field failures indicating that 46 per cent of the failure costs can be tackled by attention to only one failure type.

Defect Type	Frequency	Cost	Total Cost	% of Total
Mechanical adjustment	12	£50	£600	46
Failed printed board	6	60	360	28
Power supply unit	2	100	200	15
Miscellaneous components	15	10	150	11

9.1.4 Quality Management

(a) Organisation

The above breakdown of the activities carried out by the QA function indicates a three pronged structure as shown in figure 9.2. Depending upon

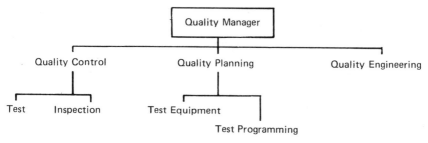

Figure 9.2

the size of the organisation quality engineering and quality planning might well be merged into one group or even a single individual.

The Quality Manager should report independently of Production and Engineering to the General Manager or Managing Director. The one disadvantage of this structure is that the Quality Manager, by having a test department, is responsible for output. This leads to a conflict between the need to achieve output targets and the responsibility to reject, and hence delay, products in test.

An alternative arrangement is for test and inspection to report to an Operations or Production Manager with the QA Manager responsible to the General Manager for Quality Engineering and Planning. This removes the disadvantage mentioned above but prevents the QA Manager controlling the test and inspection activity. There is no perfect solution to this dilemma and, in practice, a structure has to be flexible and evolve to meet the needs of the organisation.

(b) Motivation

Numerous methods have been used over the last three decades to engender a quality awareness into staff and shop floor operators in order to create a personal commitment to quality.

Zero defects campaigns involve emphasising the concept that we become tolerant to an 'acceptable' level of failure whereas no defects are acceptable. Operator involvement is enlisted to pinpoint root causes of failure. The danger with this technique, and the reason why many such attempts have failed, is that unless adequate resources are made available to progress every suggestion the scheme loses credibility and support.

Quality circles is a technique practised in Japan and now gaining popularity in the UK. It involves a small voluntary group from a defined area such as a workshop or assembly line. The group is multi-functional being made up of one or more operators, industrial engineers, quality engineers, foremen, etc. It is led by one person, not necessarily the quality engineer, who is given training in defect analysis and statistical quality control techniques (e.g. Pareto analysis). Data is gathered and analysed and the group are given a fairly free hand to recommend and implement preventive measures. A high success rate is claimed from quality circles on the basis of self motivation due to the involvement in a small group which has control over its own work area and the ability to identify tangible achievements.

(c) Reporting

It is a good discipline to report, on a monthly basis, the QA overview by means of the following three headings:

Red flag items — No more than two items otherwise the interest is diluted. Required or planned actions must be included or the item has a negative impact.

Major problems — Again actions planned must be stated. No more than four items under this heading.

Actions accomplished — These should be resolved items from problem sections of previous reports.

9.2 AUTOMATIC TEST EQUIPMENT

9.2.1 Scope and Applications

Automatic test equipment consists of:

(i) Instrumentation to provide the stimuli and make measurements.
(ii) A method of switching the various stimuli and measuring devices to each of the test points.
(iii) A method of storing the sequence of tests and measurements required.

Programmable systems have enabled (iii) to become a software item instead of a hard wired function of the circuitry. This has lead to the configuration in figure 9.3 consisting of minicomputer-controlled instrumentation.

This arrangement has the advantages of:

Diagnostic capability (alternative routes in the software algorithm).
Standard ATE (one machine for all items to be tested).

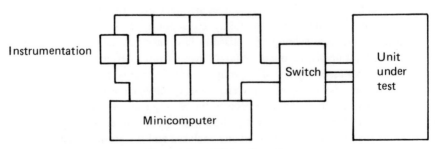

Figure 9.3   ATE configuration

Ease of extension to the instrumentation.
Speed of both test and diagnosis.
Capability for circuit analysis during design.

One option involves configuring an ATE by means of programmable instrumentation connected, via a logic interface, to a minicomputer. It is relatively cheap in terms of hardware but requires extensive programming effort to achieve diagnostic capability.

Alternatively proprietary systems are available with software packages providing a high degree of self-programming including simulation packages which enable digital test programmes to be automatically compiled from a library of known devices. This facility also enables the ATE to be used as a design aid since the performance of proposed digital circuits can be observed by means of the simulation.

Automatic test equipment is applied to:

Component testing
Printed board testing
System testing

Component testing is relatively simple since there are few test points, a restricted number of stimuli and measurements, and little requirement for diagnostic software. The other two applications require a more complex architecture to accommodate the requirements of flexibility and diagnostics.

9.2.2 Component ATE

Electronic components are divided into two groups. First are the passive components consisting of resistors, capacitors, inductors, etc., and secondly there are the active components which include all the semiconductor devices. Passive components are usually two-terminal devices presenting little test problem compared with the other group. Active semiconductors include transistors, diodes, digital and linear integrated circuits.

Both manual and programmable testers are available for transistors and diodes. The manual tester is the switch and meter type and can test for gain, leakage and saturation. Such dynamic testing although precise is time consuming — typically 5 min per component. Programmable testers usually involve thumb wheel switches or magnetic tape. The programme data is derived from the manufacturer's data sheets.

Peg board programmable test sets for digital integrated circuits are available at fairly low cost but do not compare favourably with plug in board programmed test sets. Linear integrated circuit testers must be programmed by means of an individually wired plug in card. Programming is costly but there are comparatively fewer types of linear ICs.

9.2.3 PBA Testers

(a) Component Screening   No matter how thorough component screening is carried out both manufacturing and stress related faults will occur. Printed board assembly test is therefore also required and will probably reveal:

70%	Good PBA		
30%	Faulty PBA — of which	~ 60%	Manufacturing related
		~ 40%	Component related
		?%	Small amount design related

In general all PBA ATE are configured as shown in figure 9.3. Facilities and prices vary considerably according to the range of measurements required and the level of software aids.

(b) Types of PBA Tester   PBA testers are divided into two main groups. These are Functional testers and In-circuit testers.

The Functional tester emulates the system which normally surrounds the PBA and dynamically tests it against its specification. The board is treated as a black box and its performance measured in terms of outputs and their response to inputs. Digital functional testers usually rely on the concept of a Known Good Board (KGB). Sometimes pseudo-random digital patterns are injected into the board and the performance at outputs is recorded. Bit patterns at nodes are also probed and recorded for the purpose of building up a diagnostic algorithm. Boards under test are compared with the KGB and passed or diagnosed accordingly. In more sophisticated ATE, the PBA is modelled by means of simulation software which, using a library of truth tables for the component parts, constructs output responses and nodal responses for various input conditions. Boards are tested against these simulated patterns.

Analogue testers compare stimuli and measurements at inputs, outputs and probed nodes. The KGB is again used to provide a stored memory of performance for various input conditions.

The hybrid tester combines both the facilities of analogue and digital testing at a correspondingly higher cost.

In-circuit testers are based on the philosophy of verifying that each component is within specification limits. It is then assumed that the design tolerancing is such that the PBA itself will function correctly. The in-circuit tester requires a 'Bed of Nails' to provide contact at each component node. Digital circuits are tested by determining, from a KGB, a digital bit pattern for each node. For analogue boards a nodal impedance signature is determined from a good board for d.c. and various a.c. frequencies.

(c) Programming   This is usually performed with a high-level language. One such language is ATLAS (Abbreviated Test Language for Avionic Systems) developed

and favoured by the Ministry of Defence. Some ATE systems have ATLAS compilers available. Another frequently used language is BASIC (Beginners All Purpose Symbolic Instruction Code) which in various forms is used by most ATE manufacturers under their own particular name.

Analogue programming involves a number of steps. First study the circuit and divide it into functional blocks. Specify the input conditions for a no signal quiescent state and then specify the outputs. Write a test sequence by injecting inputs according to the use of the board and measure for the correct outputs. The learning technique involves probing the board at each node surrounding a functional block. This is done in the quiescent condition and again during a functional test. The results are stored for comparison when carrying out diagnostics.

Digital programming can be lengthy and difficult due to the non-repetitive nature of the data on each highway and the often complex timing and phasing relationships in the circuit. Simulation, as mentioned above, reduces this task to a fraction. Programming entails describing the circuit as a pattern of nodes interconnected by known devices. The majority of digital devices have been described as truth tables in most of these simulation software packages.

9.2.4 Difficulties of Testing LSI

As the packaging density of LSI (Large Scale Integration) semiconductor devices increases so does the difficulty of testing printed boards containing them. The principle problems are:

Complexity — Over 5000 gates per chip which means in the order of 250 000 gates or equivalent per printed board is not uncommon.
Visibility — Reduced input/output access per function as the density increases. A common bus structure allowing single access to several functions.
Timing — Information flow is often controlled by timing signals generated within the PBA.
It is not always possible to stop in the middle of an instruction. All the data relevant to that instruction sequence has therefore to be captured and analysed if full diagnostic information is to be obtained.

Solutions to these problems require either very high speed processing within the tester to assimilate all the signals for each test step or, alternatively, some local memory at the interface with the unit under test enabling sequences to be stored for slower analysis within the ATE. The software solutions have given rise to simulation which:

Is the only realistic method with high complexity.
Reduces programming cost.
Produces automatic diagnostics.

Permits design analysis without need for a real board.
Does not require a KGB for programming.
Can predict percentage of possible fault coverage during programming.

Programming by simulation, models the KGB as already described and thereby constructs the test pattern from the software package.

9.2.5 Economics and Management of ATE

(a) Component and PBA test   The economics of board testing is influenced by the incoming component test arrangements. As PBA function density increases it becomes more important to eliminate faulty components at incoming inspection and reduce the need for expensive fault diagnosis in later stages of test. The cost of detecting and rectifying a failure can increase tenfold with each stage in the manufacturing process. Typical figures are:

Incoming test, say	10p
PBA Test, say	£1.00
System Test, say	£10.00
Field Diagnosis, say	£100.00

A 100 per cent incoming test is therefore usually desirable for LSI components although sampling is often suitable for proven discrete devices.

(b) Quantifying Costs   The following are approximate average figures at 1980 costs:

Component Testers vary between £5000 and £50 000. Programming and test times are shown in the following table.

	Programming	Test
Discrete	1 h	300–400 per h
Integrated Circuits	1 day	approx. 250 per h
	OR buy the	
	module	

PBA Testers vary between £10,000 and £200,000. Capital cost depends on a combination of frequency and digital speed, range and accuracy of instrumentation, software aids (simulation) and peripherals (teletypes, line printers, etc.).

Programming Costs should not be underestimated. The first few boards to be attempted can easily take 2 months before a satisfactory programme is written and debugged. After 6 months this should reduce to between 3 and

6 weeks per board. It must be remembered when loading and scheduling the ATE that a proportion of this time (20 per cent for digital boards and 60 per cent for analogue boards) is spent at the machine. A separate programming station will reduce this but nevertheless each programme has to be proved and debugged at the ATE using real boards. For Incircuit testing 'Bed of Nails' jigs must be designed and procured – a process taking from 4 weeks upwards and costing up to £1000 per PBA.

Testing Costs must be considered and with component screening 70 per cent of boards should pass first time taking from 1 to 3 min to test. Boards with one or more failures will take from 2 to 4 min for Incircuit testing and from 2 to 10 min for functional testing. Rework must be added to this.

Maintenance Costs are usually in the order of 8 per cent of capital and are, in most cases, covered by warranty for the first year.

Running Costs include the costs of paper for the printer, additional interface jigs and spare discs.

Existing hardwired PBA testers. It is likely that these will become spare as the newly aquired PBA tester is phased in. Such special purpose testers probably cost from £1,000 to £5,000 and involve test times from 10 to 30 min for good boards and 15–60 min for bad boards. Sometimes it is possible to utilise them after the advent of ATE by selling them to customers for second line maintenance.

(c) Justifying ATE   Any proposal must be justified on solid economic grounds rather than on the technical excellence of the equipment or on its use as a development aid. It is essential to identify ALL the costs both with and without ATE, to make assumptions which are pessimistic to the proposal, to plot the savings over at least 4 years to show when breakeven is reached, which should be less than 2 years. The following example shows the type of layout of costs which should form the backbone of a proposal. The cumulative savings are shown using only a factory overhead of 200 per cent. As a comparison the cumulative savings at the full overhead of say 600 per cent are shown in table 9.1.

(d) Utilisation and Security   A single ATE Manager is required since programming and test are not easily split and problems can too easily be referred from one to the other. In addition the allocation of machine time has to be made between those two activities. The demands will be as follows:

Production test.
Analogue programming.
Digital programming.
Design time for evaluating new circuits.
Repair of faulty boards from production.
Second line repair of boards from the field.

Table 9.1

Cost justification XYZ–ATE system

Assumptions

Current salaries with 10 per cent inflation : Factory overheads only : 1980 costs :
15k PBA p.a. : LSI early in 2nd year : All figures in £'000's.

Existing Method	Year 1	Year 2	Year 3	Year 4	Year 5
4 Testers at 120 h p.w. @ 200% @ £2.25	28	31	34	37	41
Programming and maintenance. 2 Heads.	10	11	12	13	15
LSI test equipment and time	5	5	?	?	?
Additional equipment of existing type	2	2	2	2	2
Diagnosis of returned PBA. Field and					
Line at 16 h p.w.	4	4	5	5	5
TOTAL	49	53	53	57	63

With ATE					
Capital @ £100k	38	38	38	38	–
Maintenance and supplies 8%	–	12	13	15	16
Operators (1 1st Yr.) (2 onwards)	7	16	17	19	21
Manager and programmer	12	13	15	16	18
Initial use of present testers (4 months)	10	–	–	–	–
Bureau facilities	1	–	–	–	–
Saving in system test @ 32 h p.w.					
(1 head 1st year) (2 heads onwards)	(7)	(16)	(18)	(20)	(22)
Tax payback at 52%	(52)	–	–	–	–
TOTAL	9	63	65	68	33
SAVING	40	(10)	(12)	(11)	30
Cumulative Saving	40	30	18	7	37
Cumulative saving at full overhead	71	113	157	208	306

Eventually a point will be reached where the total of these requirements exceeds
the ATE hours available. One must then consider either shift working, duplicate
ATE, where a functional tester is used with the addition of an in circuit tester for
board screening, or the purchase of a programming station.

Software security is important due to the high investment in programming
time. Data must be archived by copying on to punched tape and storing in a
separate, fireproof, area. It is also a good practice to duplicate discs since a
corrupted disc can involve lengthy reprogramming or simulation time.

10 Maintenance Handbooks

10.1 THE NEED FOR MAINTENANCE MANUALS

The main objective of a maintenance manual is to provide all the information required to carry out each maintenance task without reference to the base workshop, design authority or any other source of information. It may, therefore, include the following:

Specification of system performance and functions.
Theory of operation and usage limitations.
Method of operation.
Range of operating conditions.
Supply requirements.
Corrective and preventive maintenance routines.
Permitted modifications.
Description of spares and alternatives.
List of test equipment and its check procedure.
Disposal instructions for hazardous materials.

The manual may range from a simple card, which could hang on a wall, to a small library of information comprising many handbooks for different applications and users. Field reliability and maintainability depend, to a large degree, on the maintenance instructions. The design team, or the maintainability engineer, has to supply information to the handbook writer and to collaborate with him if the instructions are to be effective.

10.2 A TYPICAL MAINTENANCE PHILOSOPHY

Consider the provision of maintenance information for a complex system operated by a well-managed organisation with the maintenance philosophy described in section 6.1.

The system will be maintained by a permanent team (A) based on site. This team of technicians, at a fair level of competence, service a range of systems and, therefore, are not expert in any one particular equipment. Assume that the system incorporates some internal monitoring equipment and that specialised portable test gear is available for both fault diagnosis and for routine checks.

This local team carry out all the routine checks and repair most faults by means of module replacement. There is a limited local stock of some modules (LRAs) which is replenished from a central depot which serves several sites. The depot also stocks those replacement items not normally held on site.

Based at the central depot is a small staff of highly skilled specialist technicians (B) who are available to the individual sites. Available to them is further specialised test gear and also basic instruments capable of the full range of measurements and tests likely to be made. These technicians are called upon when the first line (on-site) procedures are inadequate for diagnosis or replacement. This team also visits the sites in order to carry out the more complex or critical periodic checks.

Also at the central depot is a workshop staffed with a team of craftsmen and technicians (C) who carry out the routine repairs and the checkout of modules returned from the field. The specialist team (B) is available for diagnosis and checkout whenever the (C) group are unable to repair modules.

A maintenance planning group (D) is responsible for the management of the total service operation, including cost control, coordination of reliability and maintainability statistics, system modifications, service manual updating, spares provisioning, stock control and, in some cases, a post-design service.

A preventive maintenance team (E) also based at the depot, carries out the regular replacements and adjustments to a strict schedule.

10.3 INFORMATION REQUIREMENTS FOR EACH GROUP

Group A will require detailed and precise instructions for the corrective tasks which they carry out. A brief description of overall system operation is desirable to the extent of stimulating interest but it should not be so detailed as to permit unorthodox departures from the maintenance instructions. There is little scope for initiative in this type of maintenance since speedy module diagnosis and replacement is required. Instructions for incident reporting should be included and a set format used.

Group B requires a more detailed set of data since it has to carry out fault diagnosis in the presence of intermittent, marginal or multiple faults not necessarily anticipated when the handbooks were prepared. Diagnosis should nevertheless still be to LRA level since the philosophy of first line replacement holds.

Group C will require information similar to that of Group A but concerned with the diagnosis and repair of modules. It may well be that certain repairs require the fabrication of piece parts in which case the drawings and process instructions must be available.

Group D require considerable design detail and a record of all changes. This will be essential after some years of service when the original design team may not be available to give advice. Detailed spares requirements are essential so that adequate, safe substitutions can be made in the event of a spares source or component type becoming unavailable. Consider a large population item which may have been originally subject to stringent screening for high reliability. Obtaining a further supply in a small quantity but to the same standard may be impossible and their replacement with less assured items may have to be considered. Consider also an item selected to meet a wide range of climatic conditions. A particular user may well select a cheaper replacement meeting his own conditions of environment.

Group E require detailed instructions since, again, little initiative is required. Any departure from the instructions implies a need for Group A.

10.4 TYPES OF MANUAL

(i) Preventive maintenance procedures will be listed in groups by service intervals. The service intervals can be by calendar time, switch-on time, hours flown, miles travelled, etc., as appropriate. In the same way as with calibration intervals the results and measurements at each maintenance should be used to lengthen or shorten the service interval as necessary. The maintenance procedure and reporting requirements must be very fully described so that little scope for initiative or interpretation is required. In general all field maintenance should be as routine as possible and capable of being fully described in a manual. Any complicated diagnosis should be carried out at the workshop and module replacement on site used to achieve this end. In the event of a routine maintenance check not yielding the desired result the technician should either be referred to the corrective maintenance procedure or told to replace the suspect module.

(ii) In the case of corrective maintenance (callout for failure or incident) the documentation should first list all the possible indications such as print-outs, alarms, displays, etc. Following this, routine functional checks and test point measurements can be specified. This may involve the use of a portable 'intelligent' terminal capable of injecting signals and making decisions based on the responses. A fault dictionary is a useful aid and should be continuously updated with data from the field and/or design and production areas. Full instructions should be included for isolating parts of the equipment or taking precautions where safety is involved. Precautions to prevent secondary failures being generated should be thought out by the designer and included in the maintenance procedure.

Having isolated the fault and taken any necessary precautions the next consideration is the diagnostic procedure followed by repair and checkout. Diagnostic procedures are best described in a logical flow chart. Figure 10.1 shows a segment of a typical diagnostic algorithm involving simple Yes/No decisions with paths of action for each branch. Where such a simple process is not relevant and the technician has to use initiative then the presentation of schematic diagrams and the system and circuit descriptions are important. Some faults by their nature or symptoms indicate the function which is faulty and the

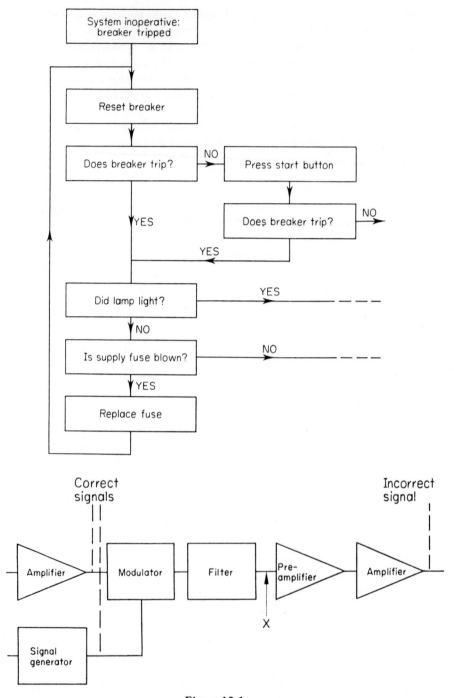

Figure 10.1

algorithm approach is most suitable. Other faults are best detected by observing the conditions existing at the interfaces between physical assemblies or functional stages. Here the location of the fault may be by a bracketing/elimination process. For example 'The required signal appears at point 12 but is not present at point 20. Does it appear at point 16? No, but it appears at point 14. Investigate unit between points 14 and 16'. The second part of figure 10.1 is an example of this type of diagnosis presented in a flow diagram. In many cases a combination of the two approaches may be necessary.

Wherever possible diagrams should be positioned opposite or adjacent to the relevant text so that frequent turning of pages backwards and forwards is minimised. In some cases an index of terminology can be helpful.

Perhaps the most highly developed technique for maintenance documentation is that known in the Royal Navy as FIMS (Functionally Identified Maintenance System). The original approach was developed by Technical Operations Inc. and applied in the United States as SIMS (Symbolic Integrated Maintenance System) and specifically in the US Navy as SIMM (Symbolic Integrated Maintenance Manual).

10.5 COMPUTER-AIDED FAULT FINDING

This involves recording information in the algorithmic form of figure 10.1 as software in a computer type store for recall and display on a VDU screen. The user/technician responds to questions on the screen by typing in simple characters from a multichoice set of questions displayed on the screen. The next question, measurement, or course of action is then presented. The advantages of this method are:

Ease of applying modifications to the software.
Ease of making copies on tape or disc.
Lengthy and complicated decision chains do not have to be on one drawing.
Only relevant information for each decision is displayed.
Ease of transporting a number of procedures on disc or tape.

A maintenance or test procedure can also be stored in a word processor. The first two of the above advantages apply and a hard copy can be produced at any time.

10.6 THE MANUAL IN PERSPECTIVE

Where maintainability is of prime importance and where the complexity of equipment and operating conditions make even the most highly skilled dependent on good documentation, this requirement reflects immediately into the design. It cannot be assumed that the provision of a handbook automatically guarantees correct diagnosis and repair. A trade-off inevitably arises between the insistence on the rigorous application of a set procedure and utilising the skill of the technician. Unless the handbook is clear and concise it will almost certainly be cast aside in favour of initiative and trial and error as a result of which the repair time will rise accordingly.

As stated at the beginning the handbook may range from a card to a library of information. Before despising the humble card consider that, put to good use, it is more valuable than a dozen volumes which are consulted in desperation only after the equipment has been rendered nearly unserviceable by trial and error.

11 Making Use of Field Feedback

Failure data can be collected from prototype and production models or from the field. In either case a formal failure reporting document is necessary in order to ensure that the feedback is both consistent and adequate. Field information is far more valuable since it concerns failures and repair actions which have taken place under real conditions. Since recording field incidents relies on people it is subject to errors, omissions and misinterpretation. It is therefore important to collect all field data using a formal document. Information of this type has a number of uses the main two being feedback, resulting in modifications to prevent further defects, and the acquisition of statistical reliability and repair data. In detail then:

They indicate design and manufacture deficiencies and can be used to support reliability growth programmes.
They provide quality and reliability trends.
They provide subcontractor ratings.
They contribute statistical data for future reliability and repair time predictions.
They assist second line maintenance (workshop).
They enable spares provisioning to be refined.
They enable routine maintenance intervals to be revised.
They enable the field element of quality costs to be identified.

A failure reporting system should be established for every project and product. Customer cooperation with a reporting system is essential if feedback from the field is required and this could well be sought, at the contract stage, in return for some other concession.

11.2 INFORMATION TO BE RECORDED

The failure report form must demand information covering the following:

Repair Time — active and passive
Type of Fault — primary or secondary, random or induced, etc.

Nature of Fault — open or short circuit, drift condition, wearout, etc.

Fault Location — exact position and detail of LRA or component.

Environmental conditions — where these are variable record conditions at time of fault if possible.

Action taken — exact nature of replacement or repair.

Personnel involved.

Equipment used.

Spares used.

11.3 DIFFICULTIES INVOLVED

The main problems associated with failure recording are:

MOTIVATION: If the field service engineer can see no purpose in recording information it is likely that items will be either omitted or incorrectly recorded. The purpose of fault reporting and the ways in which it can be used to simplify his task need to be explained. If he is frustrated by unrealistic time standards, poor working conditions and inadequate instructions then the failure report is the first task which he will skimp or omit. A regular circulation of field data summaries to the field engineers is the best, possibly the only, way of encouraging feedback. It will help him to see the overall field picture and advice on diagnosing the more awkward faults will be appreciated.

VERIFICATION: Once the failure report has left the person who completes it the possibility of subsequent checking is remote. If repair times or diagnoses are suspect then it is likely that they will go undetected or unverified. Where failure data are obtained from customer's staff the possibility of challenging information becomes even more remote.

COST: Failure reporting is costly both in terms of the time to complete failure report forms and in terms of the manhours of interpretation of the information. For this reason both supplier and customer are often reluctant to agree to a comprehensive reporting system. If the information is correctly interpreted and design or manufacturing action taken to remove failure sources then the cost of the activity is likely to be offset by the savings and the idea must be 'sold' on this basis.

RECORDING NON-FAILURES: The situation arises where a failure is recorded although none exists. This can occur in two ways. Firstly there is the habit of locating faults by replacing suspect but not necessarily failed components. When the fault disappears the first, wrongly removed, component is not replaced and is hence recorded as a failure. Failure rate data are therefore artificially inflated and spares depleted. Secondly there is the interpretation of secondary failures as primary failures. A failed component may cause stress conditions upon another which may, as a result, fail. Diagnosis may reveal both failures but not always which one occurred first. Again failure rates become wrongly inflated. More complex maintenance instructions and the use of higher grade personnel will help reduce these problems at a cost.

11.4 ANALYSIS AND PRESENTATION OF RESULTS

Once collected, data must be analysed and put to use or the system of collection will lose credibility and, in any case, the cost will have been totally wasted. In section 9.1.3e the Pareto method of presenting failure information was shown and a more complete example of three months failure data is given in figure 11.1. Note the emphasis on cost and that the total has been shown as a percentage of sales. It is clear that engineering effort could profitably be directed at the first two items which together account for 38 per cent of the failure cost. The first item is a mechanical design problem and the second a question of circuit tolerancing.

It is also important to know whether the failure rate of a particular failure type is increasing, decreasing or constant. This will influence the engineering response. A decreasing failure rate indicates the need for further action in test to eliminate the early failures. Increasing failure rate shows wearout requiring either a design solution or preventive replacement. Constant failure rate suggests a reliability level which is inherent to that design configuration. Chapter 13

---

1. Summary of Data

Number of machines in field	50	
Operating hours (this period)	5320	
Number of corrective calls	39	
Total cost of calls	£4250	labour, travel and spares
Total cost as % of sales	4%	

2. Incident Analysis

Repetitive Failures	Frequency	Cost £	% of total
a) Mechanical transporter assembly — belt adjustment	4	935	22
b) Receiver carrier detector drift	9	680	16
c) Electromechanical relays	4	340	8
d) Gear Meshing	3	340	8
e) Printed Board 182c output VT2	2	300	7
f) Lamps	2	170	4
Non-repetitive Faults			
g) Printed Board 424a IC5 h) Printed Board 111e R2	15	1485	35
etc			
	39	4250	100

Figure 11.1   Quarterly incident report summary – product Y

explains how failure data can be analysed to quantify these trends. The report in figure 11.1 might well contain other sections showing reliability growth, analysis of wearout, progress on engineering actions since the previous report, etc.

11.5 EXAMPLES OF FAILURE REPORT FORMS

Figure 11.2 shows an example of a well-designed and thorough failure recording form as used by the European Companies of the International Telephone and Telegraph Corporation. This single form strikes a balance between the need for detailed failure information and the requirement for a simple reporting format. A feature of the ITT form is the use of four identical print through forms. The information is therefore accurately recorded four times with minimum effort.

As an example of the need for more elaborate reporting, consider an air traffic control service which is essentially concerned with the safety of life in a dynamic situation. It is not surprising therefore to find a detailed maintenance reporting system in such an organisation. Three of the forms used by the British Civil Aviation Authority are shown in figure 11.3. They deal with corrective and planned maintenance reporting in great detail.

It is unfortunate that few forms give adequate breakdown of maintenance times separated into the various passive and active elements. To identify and record this level of information increases the maintenance time and cost. It has to be justified if a special investigation is required. Such an analysis can result in improved maintenance procedures in which case it may pay for itself by reducing long term costs.

	System	Sub-system	Module/ sub-assembly	

ITTE Failure Report and Action Form

Report number:
Report date:
Report completed by:
Company:

Type			
Serial number			
Location/identification			
On-time (cumulative)			
Down time (this failure)			
Active repair time			

To be completed on site

System status:
    Field service ☐
    Field trial ☐
    Production prototype ☐
    Model ☐

Effect of failure on system:
    Complete system failure ☐
    Major degradation ☐
    Minor degradation ☐
    None ☐

On-site diagnosis:
    No defect found ☐
    Part failure ☐
    Installation defect ☐
    Manufacturing defect ☐
    Design defect ☐
    Program defect ☐
    On-site human error ☐
    Other ☐

Action taken:
    Replace module ☐
    Repair ☐
    Modification ☐
    Program reload ☐
    Other ☐

Details of symptoms, diagnosis and failure:

Details of action taken:

To be completed at Designated Centre

Project engineering action:	Name	Company/Dept	Signature	Date completed
− Consolidate with filed data				
− For immediate analysis/action by:				
Engineering				
Manufacturing				
Quality assurance				
Purchasing				
Other				

Analysis and action taken:

Engineering change no.:
Dated:
Follow-up report. Ref no.:
Dated:
Name:
Signature:
Date:

For information to:

Figure 11.2   ITT Europe failure report and action form

CA 1520 September 1972	NATIONAL AIR TRAFFIC SERVICE	EQUIPMENT DEFECT REPORT 97

A	FOR USE OF ADP STAFF ONLY	LOCATION	PREFIX	SERIAL NUMBER	DATE TIME WORK STARTS DAY MONTH TIME
%	6	1	2	3	4

FOUND BY PLANNED MAINT.?	REPAIR TIME HOURS MINS.	MAN HOURS HOURS MINS.	SYSTEM CONTROL REFERENCE LOCATION PREFIX NUMBER LINE	EQUIPMENT
5 Y N	6	7	8	9

	DESIGNATION	ASSEMBLY	SUB ASSEMBLY	ITEM ACTIONED	ACTION	ACTION CODES		FOR STATION USE
10						Replaced, Open Circuit	A	REPAIR TIME
11						Replaced, Short Circuit	B	
12						Replaced, Intermittent	C	
13						Replaced, Over Heated	D	
14						Replaced, Out of Tolerance	E	
15						Replaced, Other Reason	F	
16						Adjusted	G	
17						Repaired	H	
18						Wiring Fault Repaired	J	
19						Cleaned	K	
20						Item Dried Out	L	
21						Lubricated	M	
22						Rubber Moulding Perished	N	
23						No Fault Found	W	MAN HOURS
						Refurbished	X	
						Sent to Industry	Y	
						Scrapped	Z	
						Reset	R	
						Programme	P	
						CHARACTERS		
						Letter I — I — Figure One	I	
						Letter O — O — Figure Nought	Ø	
						CROSS REFERENCE		

B	SPECIAL INVESTIGATION SECTION			
✱ 24	25	26	27	28
✱ 29	30	31	32	33

C SYMPTOMS OF FAULT OR WORK TO BE DONE

Name and Grade in Block Capitals .... *Date* ....

D COMMENTS OR WORK CARRIED OUT

FACILITY	
A	
ITEM SERIAL NUMBER	
B	
LOCATION OF EQUIPMENT	HOURS RUN
C	D

Name and Grade in Block Letters .... *Date* ....

Figure 11.3a NATS equipment defect report

CA 1519 September 1972	NATIONAL AIR TRAFFIC SERVICE	PLANNED MAINTENANCE REPORT

A — ADP REPORTING SECTION

FOR USE OF ADP STAFF ONLY	LOCATION	PREFIX	SERIAL NUMBER	TIME MAINTENANCE STARTS
% **7**	1	2	3	DAY MONTH TIME 4

REASON FOR MAIN- TENANCE	MAINTENANCE TIME HOURS MINS	MAN HOURS HOURS MINS	SYSTEM CONTROL REFERENCE LOCATION PREFIX NUMBER LINE	EQUIPMENT
5 P 6	7	8	9	

B — SPECIAL INVESTIGATION SECTION

10	11	12	13	14
15	16	17	18	19

C — PLANNED MAINTENANCE WORK SHEET

HAS USER BEEN INFORMED?	DATE TIME MAINTENANCE STARTS (PLANNED) Day Month Time	DATE TIME OF SERVICE INTERRUPTION (PLANNED) Day Month Time	SERVICE DOWN TIME (PLANNED) Hrs. Mins.	TOTAL WORKING TIME (PLANNED) Hrs. Mins.
C1 Y NA	C2	C3	C4	C5

Equipment Designation	Work Planned	Work Com- pleted	Equipment Adjusted	Maint. Time	Man Hours	DETAILS	E.D.R. Serial Number

DISTRIBUTION
Tels HQ .... Green
Equipment Records ......Black
As instructed by
C.T.O.I.s ....Blue

CHARACTERS
Figure One I
Figure Nought ... 0
Letter I............I
Letter O..............O

REASON FOR MAINTENANCE CODES
Planned Preventive ....PA   Modification by Station Staff.. PD   Engineering Investigation ...... PG
Replacement of a Unit ......PC   Independent Inspection............ PE
Engineering Modification .... PF

Name and Grade in Block Capitals				Date
Checked by	S.T.O.	C.T.O.	Cross Reference	

Figure 11.3b   NATS planned maintenance report

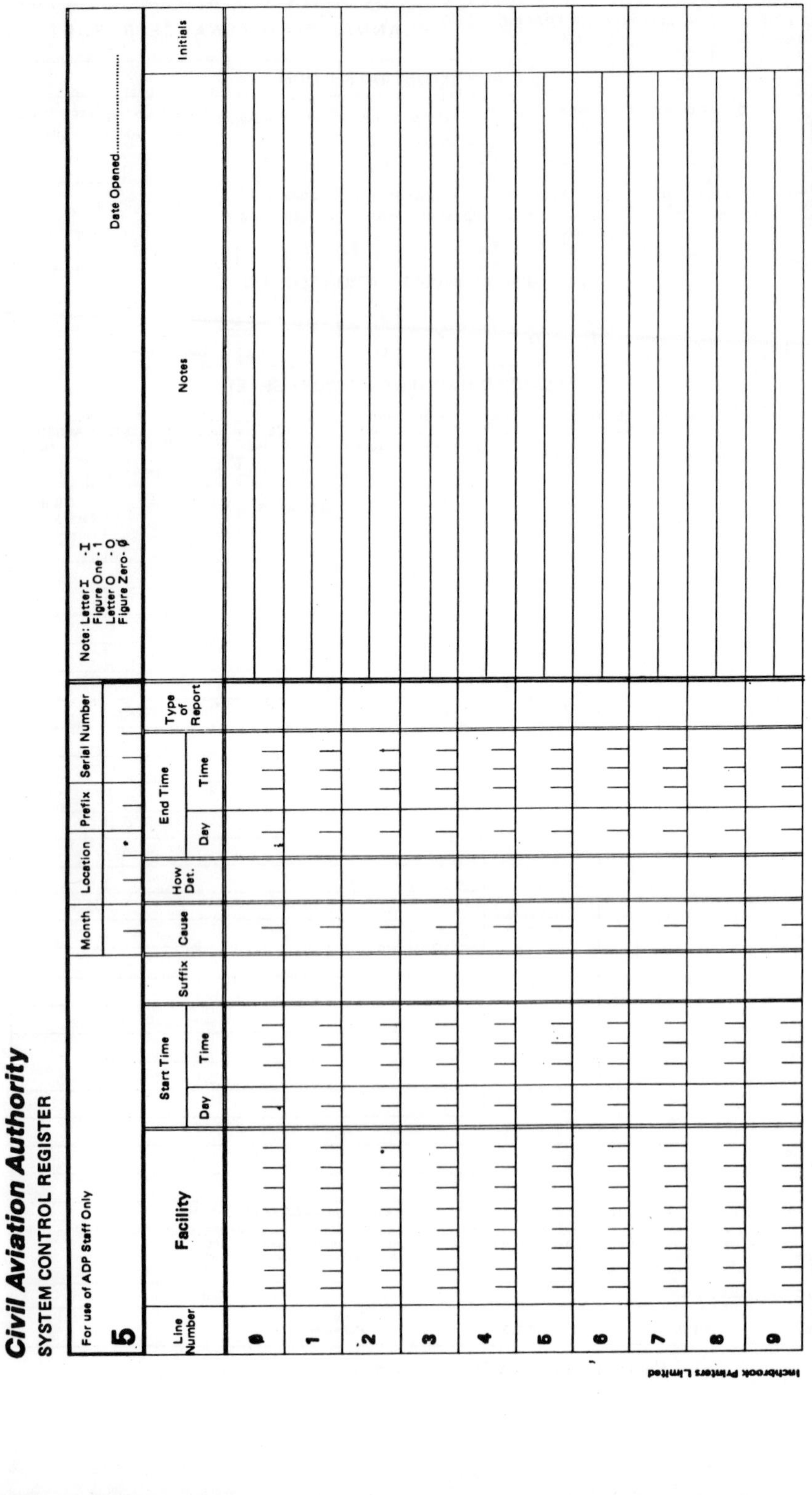

Civil Aviation Authority
SYSTEM CONTROL REGISTER

For use of ADP Staff Only

5

Date Opened................

Note: Letter I - I
Figure One - 1
Letter O - O
Figure Zero - Ø

Line Number	Facility	Start Time		Suffix	Cause	How Det.	End Time		Prefix	Serial Number	Type of Report	Notes	Initials
		Day	Time				Day	Time					
0													
1													
2													
3													
4													
5													
6													
7													
8													
9													

Month   Location   Prefix   Serial Number

CA Form 1518
230775

Inchbrook Printers Limited

Distribution: Green Copy - ADP
Black Copy - Station Copy
Blue Copy - C.T.O.

Figure 11.3c   CAA system control register

Part III

Making Measurements and Predictions

12 Interpreting Data and Demonstrating Reliability

This chapter deals with the interpretation of failure rates and MTBFs for the special case where random failures are assumed. We are dealing thus with constant failure rates and the equality $\lambda = 1/\theta$ applies. The next chapter will explore the analysis of variable failure rates.

12.1 INFERENCE AND CONFIDENCE LEVELS

In section 3.2 the concept of a point estimate of failure rate (λ) or MTBF (θ) was introduced. In the model N items showed k failures in T cumulative hours and the observed MTBF $(\hat{\theta})$ of that sample measurement was T/k. If the test were repeated, and another value of T/k obtained, it would not be exactly the same as the first and, indeed, a number of tests would yield a number of values of MTBF. Since these estimates are the result of sampling they are called point estimates and have the sign $\hat{\theta}$. It is the true MTBF of the batch which is required and the only way to obtain it is to allow the entire batch to fail and then to evaluate T/k. This compares with the theoretical determination of MTBF in equation (3.5) of section 3.5 where it was seen that

$$\text{MTBF} = \int_0^\infty \frac{N_s(t)}{N}\,dt$$

The limits of integration are consistent with the statement that all devices must fail if the true MTBF is to be determined. Such a test will, of course, yield accurate data but, alas, no products at the end. In practice we are forced to truncate tests after either a given number of hours or failures. The first is called a time-truncated test and the second a failure-truncated test. The problem to be solved is that a statement about MTBF is required when only a sample is available.

The process of making a statement about a population of items based on the evidence of a sample is known as statistical inference. It involves, however, the additional concept of confidence level. This is best illustrated by means of an example. Figure 12.1 shows a distribution of heights of a group of people in histogram form. Superimposed onto the histogram is a curve of the Normal

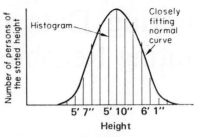

Figure 12.1   Distribution of heights

distribution. The practice in statistical inference is to select a mathematical distribution which closely fits the data. Statements based on the distribution are then assumed to apply to the data.

In the figure there is a good fit between the normal curve, having a mean of 5′10″ and a standard deviation (measure of spread) of 1″, and the heights of the group in question. Consider, now, a person drawn, at random, from the group. It is permissible to predict, from a knowledge of the normal distribution, that the person will be 5′10″ tall or more providing that it is stated that the prediction is made with 50 per cent confidence. This really means that we anticipate being correct 50 per cent of the time if we continue to repeat the experiment. On this basis an infinite number of possible heights can be predicted providing that an appropriate confidence level accompanies each. For example:

$$5′11″ \text{ or more at } 15.9 \text{ per cent confidence}$$
$$6′\ 0″ \text{ or more at } \ \ 2.3 \text{ per cent confidence}$$
$$6′\ 1″ \text{ or more at } \ \ 0.1 \text{ per cent confidence}$$
$$\text{OR between } 5′9″ \text{ and } 5′11″ \quad \text{at} \qquad 68.2 \text{ per cent confidence}$$

The inferred measurement and the confidence level can, hence, be traded off against each other.

12.2 THE χ^2 TEST

Returning to the estimates of MTBF it is possible to employ the same technique of stating an MTBF together with a confidence level if the way in which the values are distributed is known. It has been shown that the expression

$$\frac{2k\hat{\theta}}{\theta} \text{ (random failures assumed)}$$

follows a χ^2 distribution with $2k$ degrees of freedom, where the test is truncated at the kth failure.

We know already that

$$\hat{\theta} = \frac{T}{k} = \frac{\text{Accumulated test hours}}{\text{Number of failures}}$$

Therefore

$$\frac{2k\hat{\theta}}{\theta} = \frac{2kT}{k\theta} = \frac{2T}{\theta}$$

so that $2T/\theta$ is χ^2 distributed.

If a value of χ^2 can be fixed for a particular test then $2T/\theta$, and hence θ, can be stated to lie between specified confidence limits. In practice the upper limit is usually set at infinity and one talks of an MTBF of some value or greater. This is known as the single sided lower confidence limit of MTBF. Figure 12.2 shows a distribution of χ^2. The area of the shaded portion is the probability of χ^2 exceeding that particular value at random.

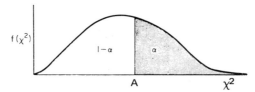

Figure 12.2   Single-sided confidence limits

In order to fix a value of χ^2 it is necessary to specify two parameters. The first is the number of degrees of freedom (twice the number of failures) and the second is the confidence level. The tables of χ^2 at the end of this book have columns and rows labelled α and n. α is the confidence level of the χ^2 distribution and n is the number of degrees of freedom. The limits of MTBF, however, are required between some value, A, and infinity. Since $\theta = 2T/\chi^2$ the value of χ^2 corresponding to infinite θ is zero. The limits are therefore zero and A. In figure 12.2 if α is the area to the right of A then $1 - \alpha$ must be the confidence level of θ.

If the confidence limit is to be at 60 per cent, the lower single sided limit would be that value which the MTBF exceeds 6 times out of 10. Since the degrees of freedom can be obtained from $2k$ and $\alpha = (1 - 0.6) = 0.4$ then a value of χ^2 can be obtained from the tables.

From $2T/\chi^2$ it is now possible to state a value of MTBF at 60 per cent confidence. In other words such a value of MTBF or better would be observed 60 per cent of the time. It is written $\theta_{60\%}$.

In a replacement test (each failed device is replaced immediately) 100 devices are tested for 1000 h during which 3 failures occur. The third failure occurs at 1000 h at which point the test is truncated. We shall now calculate the MTBF of the batch at 90 per cent and 60 per cent confidence levels.

1. Since this is a replacement test T is obtained from the number under test multiplied by the linear test time. Therefore $T = 100\ 000$ h and $k = 3$.
2. Let $n = 2k = 6$ degrees of freedom. For 90 per cent confidence $\alpha = (1 - 0.9) = 0.1$ and for 60 per cent confidence $\alpha = 1 - 0.6 = 0.4$.

3. Read off χ^2 values of 10.6 and 6.21.(see p. 219)
4. $\theta_{90\%} = 2 \times 100\,000/10.6 = 18\,900$ h.
   $\theta_{60\%} = 2 \times 100\,000/6.21 = 32\,200$ h.

Compare these results with the original point estimate of $T/k = 100\,000/3 = 33\,333$ h. It is possible to work backwards and discover what confidence level is actually applicable to this estimate. $\chi^2 = 2T/\theta = 200\,000/33\,333 = 6$. Since n is also equal to 6 it is possible to consult the tables and see that this occurs for a value of α slightly greater than 0.4. The confidence with which the MTBF may be quoted as 33 333 h is therefore less than 60 per cent. It cannot be assumed that all point estimates will yield this value and, in any case, a proper calculation, as outlined, should be made.

In the above example the test was failure truncated. For a time-truncated test one must be added to the number of failures (2 to the degrees of freedom) for the lower limit of MTBF. This takes account of the possibility that, had the test continued for a few more seconds, a failure might have occurred. In the above single-sided test the upper limit is infinity and the value of MTBF is, hence, the lower limit. A test with zero failures can now be interpreted. Consider 100 components for 50 h with no failures. At a 60 per cent confidence we have $\theta_{60\%} = 2T/\chi^2 = 2 \times 50 \times 100/\chi^2$. Since we now have $\alpha = 0.4$ and $n = 2(k+1) = 2$, $\chi^2 = 1.83$ and $\theta = 20\,000/1.83 = 11\,900$ h. Suppose that an MTBF of 20 000 h was required. The confidence with which it has been proved at this point is calculated as before. $\chi^2 = 2T/\theta = 20\,000/20\,000 = 1$. This occurs at $\alpha = 0.6$, therefore the confidence stands at 40 per cent. If no failures occur then, as the test continues, the rise in confidence can be computed and observed. Furthermore the length of the test (for zero failures) can be calculated in advance for a given MTBF and confidence level.

12.3 DOUBLE-SIDED CONFIDENCE LIMITS

So far lower single-sided statements of MTBF have been made. Sometimes it is required to state that the MTBF lies between two confidence limits. Once again $\alpha = (1 - \text{confidence level})$ and is split equally on either side of the limits as shown in figure 12.3.

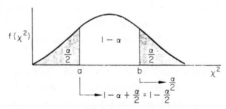

Figure 12.3   Double-sided confidence limits

The two values of χ^2 are found by using the tables twice, firstly at $n = 2k$ and at $1 - \alpha/2$ (this gives the lower limit of χ^2) and secondly at $n = 2k$ ($2k + 2$ for time truncated) and at $\alpha/2$ (this gives the upper limit of χ^2). Once again the upper limit of χ^2 corresponds with the lower limit of MTBF and vice versa. Figure 12.3 shows how $\alpha/2$ and $1 - \alpha/2$ are used. The probabilities of χ^2 exceeding the limits are the areas to the right of each limit and the tables are given accordingly.

Each of the two values of χ^2 can be used to obtain the limits of MTBF from the expression $\theta = 2T/\chi^2$. Using the earlier example, assume that the upper and lower limits of MTBF for a 90 per cent confidence band are required. In other words limits of MTBF are required such that 90 per cent of the time it will fall within them. $T = 100\,000$ h and $k = 3$. The two values of χ^2 are obtained:

$$n = 6, \alpha = 0.95, \chi^2 = 1.64$$

$$n = 6, \alpha = 0.05, \chi^2 = 12.6$$

This yields the two values of MTBF — 15 900 h and 122 000 h, in the usual manner, from the expression $\theta = 2T/\chi^2$.

Hence the MTBF lies between 15 900 and 122 000 h with a confidence of 90 per cent.

12.4 SUMMARISING THE χ^2 TEST

The following list of steps summarises the use of the χ^2 tables for interpreting the results of reliability tests.

1. Measure T (accumulated test hours) and k (number of failures).
2. Select a confidence level and let $\alpha = (1 - \text{confidence level})$.
3. Let $n = 2k$ ($2k + 2$ for lower limit MTBF in time-truncated test).
4. Note the value of χ^2 from the tables.
5. Let MTBF at the given confidence be $2T/\chi^2$.
6. For double-sided limits use the above procedure twice at

$n = 2k : 1 - \alpha/2$ (upper limit of MTBF)
$n = 2k$ ($2k + 2$) : $\alpha/2$ (lower limit of MTBF)

It should be noted that, for constant failure rate conditions, 100 components under test for 20 h yield the same number of accumulated test hours as 10 components for 200 h. Other methods of converting test data into statements of MTBF are available but the χ^2 distribution method is the most flexible and easy to apply. MTBFs are computed usually at the 60 per cent and 90 per cent confidence levels.

12.5 RELIABILITY DEMONSTRATION

Imagine that, as a manufacturer, you have evaluated the MTBF of your components at some confidence level using the techniques outlined, and that you have sold them to me on the basis of such a test. I may well return, after some time, and say that the number of failures experienced in a given number of hours yields a lower MTBF, at the same confidence, than did your earlier test. You could then suggest that I wait another month by which time there is a chance that the number of failures and the number of test hours will have swung the calculation in your favour. Since this is hardly a suitable way of doing business it is necessary for consumer and producer to agree on a mutually acceptable test for accepting or rejecting batches of items. Once the test has been passed there is to be no question of later rejection on discovering that the batch passed on the strength of an optimistic sample. On the other hand there is no redress if the batch is rejected, although otherwise acceptable, on the basis of a pessimistic sample. The risk that the batch, although within specification, will fail due to a pessimistic sample being drawn is known as the producer's risk and has the symbol α (not to be confused with the α in the previous section). The risk that a 'bad' batch will be accepted due to an optimistic sample is known as the consumer's risk, β. The test consists of accumulating a given number of test hours and then accepting or rejecting the batch on the basis of whether or not a certain number of failures have been observed.

Imagine such a test where the sample has to accumulate T test hours with no failures in order to pass. If the failure rate is assumed to be constant with a value λ then the probability of observing no failures in T test hours is $e^{-\lambda T}$ (from the Poisson distribution). Such a zero failures test is represented in figure 12.4 which is a graph of the probability of observing no failures (in other words of passing the test) against the anticipated number of failures given by λT. This type of test is known as a Fixed Time Demonstration Test. It can be seen from the graph that, as the failure rate increases, the probability of passing the test falls.

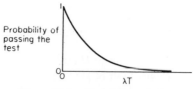

Figure 12.4   Zero failures test

The problem with this type of testing rests with the degree of DISCRIMINATION in each test. In order to understand discrimination consider the following extreme example:

A component has an acceptable failure rate of 300×10^{-9}/h (N.B. approx 1 in

380 yr). 50 are tested for 1000 h (approx $5\frac{1}{2}$ years of test). λT is therefore $\frac{5.5}{380} = 0.014$ and the probability of passing the test is $e^{-0.014} = 98.6$ per cent.

Suppose that a second test is made from a batch whose failure rate is three times the first batch (i.e. 900×10^{-9}/h).

Now the probability of passing the test is $e^{-\lambda T} = e^{-0.043} = 95.8$ per cent. Whereas the acceptable batch is 98.6 per cent sure of acceptance ($\alpha = 1.4$ per cent) the 'bad' batch is only 4.2 per cent sure of rejection ($\beta = 95.8$ per cent). In other words although the test is satisfactory for passing batches of the required failure rate it is a poor discriminator whose acceptance probability does not fall quickly as the failure rate increases.

A test is required which not only passes acceptable batches (a sensible producer's risk would be between 5 and 15 per cent) but rejects batches with a significantly higher failure rate. Three times the failure rate should reduce the acceptance probability to 15 per cent or less. The only way that this can be achieved is to increase the test time so that the acceptance criterion is much higher than zero failures.

In general the criterion for passing the test is n or fewer failures and the probability of passing the test is:

$$P_{0-n} = \sum_{i=0}^{n} \frac{\lambda^i T^i e^{-\lambda T}}{i!}$$

This expression yields the family of curves shown in figure 12.5 which includes the special case ($n = 0$) of figure 12.4. These curves are known as Operating Characteristics (O.C. Curves), each one representing a test plan.

Each of these curves represents a valid test plan and to demonstrate a given failure rate there is a choice of $0, 1, 2, 3, \ldots n$ failure criterion tests with corresponding values of T. The higher the number of failures the greater the number of test hours are required. Figure 12.6 shows the improvement in discrimination as n increases. Note that n is replaced by c which is the usual convention. The extreme case where everything is allowed to fail and c equals

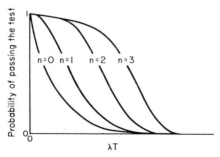

Figure 12.5   Family of O.C. curves

the population is shown. Since there is no question of uncertainty under these circumstances the probability of passing the test is either one or zero depending upon the batch failure rate. The question of sampling risks does not arise.

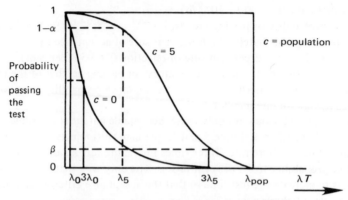

Figure 12.6   O.C curves showing discrimination

Consider the $c = 0$ plan and note that a change from λ_0 to $3\lambda_0$ produces little decrease in the acceptance probability and hence a poor consumer's risk. If the consumer's risk were to be 10 per cent the actual failure rate would be a long way to the right on the horizontal axis and would be many times λ_0. This ratio is known as the Reliability Design Index or Discrimination Ratio. Looking, now, at the $c = 5$ curve, both producer and consumer risks are reasonable for a $3:1$ change in failure rate. In the extreme case of 100 per cent failures both risks reduce to zero.

Figure 12.7 is a set of Cumulative Poisson Curves which enable the test plans and risks to be evaluated as in the following example:

A failure rate of 3×10^{-4}/h is to be demonstrated using ten items. Calculate the number of test hours required if the test is to be passed with 4 or less failures and the probability of rejecting acceptable items (α) is to be 10 per cent.

1. Probability of passing test $= 1 - 0.1 = 0.9$.
2. Using figure 12.7 the corresponding value for $c = 4$ at 0.9 is 2.45.
3. $\lambda T = 3 \times 10^{-4} \times T = 2.45$.
   Therefore $T = 8170$ h.
4. Since there are 10 items the test must last 817 h with no more than 4 failures.

If the failure rate is 3 times the acceptable value calculate the consumer's risk, β.

1. $3\lambda T = 3 \times 3 \times 10^{-4} \times 8170 = 7.35$.
2. Using figure 12.7 for $m = 7.35$ and $c = 4 : P_{0-4} = 0.15$.
3. The consumer's risk is therefore 15 per cent.

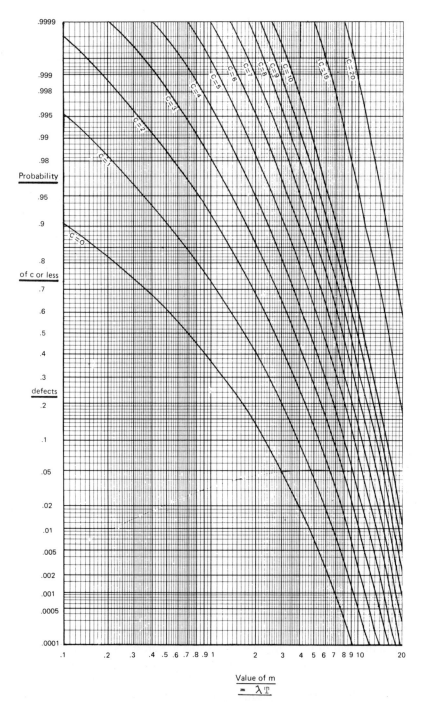

Figure 12.7   Poisson curves

The reader might care to repeat this example for a zero failures test and verify for himself that, although T is as little as 333 h, β rises dramatically to 74 per cent. The difficulty of high reliability testing can now be appreciated. For example an equipment which should have a one year MTBF requires at least 3 years of testing to demonstrate its MTBF with acceptable risks. If only one item is available for test then the duration of the demonstration would be 3 years. In practice far larger MTBFs are aimed for, particularly with submarine and satellite systems, and demonstration testing as described in this chapter is not applicable.

12.6 SEQUENTIAL TESTING

The above type of test is known as a Fixed-Time Demonstration. Due to the difficulties of discrimination any method that results in a saving of accumulated test hours without changing any of the other parameters is to be welcomed.

Experience shows that the Sequential Demonstration Test tends to achieve results slightly faster than the equivalent fixed time test. Figure 12.8 shows how a sequential reliability test is operated. Two parallel lines are constructed so as to mark the boundaries of the 3 areas — Accept, Reject and Continue Testing. As test hours are accumulated the test proceeds along the x-axis and as failures occur the line is moved vertically one unit per failure. Should the test line cross the upper boundary too many failures have been accrued for the hours accumulated and the test has failed. If, on the other hand, the test crosses the lower boundary sufficient test hours have been accumulated for the number of failures and the test has passed. As long as the test line remains between the boundaries the test must continue.

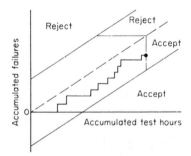

Figure 12.8   Truncated sequential demonstration test

Should a time limit be set to the testing then a truncating line is drawn as shown to the right of the diagram so that if the line crosses above the midpoint the test has failed. If, as shown, it crosses below the midpoint then the test has passed. If a decision is made by crossing the truncating line rather than one of the boundary lines then the consumer and producer risks calculated for the test no longer apply and must be recalculated.

As in the fixed time test the consumer's risk, producer's risk and the MTBF associated with each are fixed. The ratio of the two MTBFs (or failure rates) is the reliability design index. The lines are constructed from the following equations:

$$y_{upper} = \frac{\frac{1}{\theta_1} - \frac{1}{\theta_0}}{\log_e \frac{\theta_0}{\theta_1}} T + \frac{\log_e A}{\log_e \frac{\theta_0}{\theta_1}} : A \approx \frac{1 - \beta}{\alpha} \text{ and } \frac{\beta}{1 - \alpha}$$

provided α and β are small (less than 25 per cent).

The equation for y_{lower} is the same with $\log_e B$ substituted for $\log_e A$. If the risks are reduced then the lines move further apart and the test will take longer. If the design index is reduced, bringing the two MTBFs closer together, then the lines will be less steep making it harder to pass the test.

12.7 SETTING UP DEMONSTRATION TESTS

In order to conduct a *demonstration test* (sometimes called a verification test) the following conditions, in addition to the statistical plans already discussed, must be specified.

1. Values of consumer's risk and acceptable MTBF. The manufacturer will then decide on his risk and upon a reliability design index. This has already been examined in this chapter. A failure distribution must be agreed (this chapter has only dealt with random failures). A test plan can then be specified.
2. The sampling procedure must be defined in terms of sample size and from where and how the samples should be drawn.
3. Both environmental and operational test conditions must be fixed. This includes specifying the location of the test and the test personnel.
4. Failure must be defined so that there will be no argument over what constitutes a failure once the test has commenced. Exceptions should also be defined, i.e. failures which are to be disregarded (failures due to faulty test equipment, wrong test procedures, etc.).
5. If a 'burn-in' period is to be allowed, in order that early failures may be disregarded, this too must be specified.

The emphasis in this chapter has been on component testing and demonstration, but if equipment or systems are to be demonstrated, the following conditions also must be specified:

1. Permissible corrective or preventive maintenance during the test (e.g. replacement of parts before wearout, routine care).

2. Relevance of secondary failures (failures due to fluctuations in stress caused by other failures).
3. How test time relates to real time. (24 h operation of a system may only involve 3 h of operation of a particular unit).
4. Maximum setting-up and adjustment time permitted before the test commences.

U.S. MILITARY STANDARD 781B — Reliability Testing, Exponential Distribution contains both fixed time and sequential test plans. Alternatively plans can be constructed from the equations and curves given in this chapter.

EXERCISES

1. A replacement test involving 50 devices is run for 100 hours and then truncated. Calculate the MTBF (single sided lower limit) at 60 per cent confidence:

   (a) If there are 2 failures.
   (b) If there are zero failures.

2. The items in Exercise 1 are required to show an MTBF of 5000 h at 90 per cent confidence. What would be the duration of the test, with no failures, to demonstrate this?

3. The producer's risk in a particular demonstration test is set at 15 per cent. How many hours must be accumulated, with no failures, to demonstrate an MTBF of 1000 h? What is the result if a batch is submitted to the test with an MTBF of 500 h? If the test were increased to 5 failures what would be the effect on T and β?

13 Interpreting Variable Failure Rate Data

13.1 THE WEIBULL DISTRIBUTION

The bathtub curve in figure 3.2 showed that, in addition to random failures, there are distributions of increasing and decreasing failure rate. In these variable failure rate cases it is of little value to consider the actual failure rate since only Reliability and MTBF are meaningful. In chapter 3 we saw that:

$$R(t) = \exp\left[-\int_0^t \lambda(t)\, dt\right]$$

Since the relationship between failure rate and time takes many forms, and depends on the device in question, the integral cannot be evaluated for the general case. Even if the variation of failure rate with time were known it might well be of such a complicated nature that the integration would prove far from simple. In practice it is found that the relationship can usually be described by the following three-parameter distribution known as the Weibull Distribution.

$$R(t) = \exp\left[-\left(\frac{t-\gamma}{\eta}\right)^\beta\right]$$

In the constant failure rate case it was seen that statements of MTBF and reliability could be made from the failure rate parameter which completely defined the distribution. In the Weibull case the reliability function requires three parameters (γ, β, η). They do not have a physical meaning as does failure rate and must be treated as merely numbers which allow us to compute reliability and MTBF. In the special case of $\gamma = 0$ and $\beta = 1$ the expression reduces to the simple exponential case with $\eta = $ MTBF. This is slightly misleading because in the general case η is not equal to MTBF.

The Weibull expression can be reduced to a straight line equation by taking logarithms twice:

If $\qquad\qquad 1 - R(t) = Q(t) \ldots$ the unreliability
$\qquad\qquad\qquad\qquad\qquad\qquad$ (probability of failure in t)

Then

$$1 - Q(t) = \exp\left[-\left(\frac{t - \gamma}{\eta}\right)^{\beta}\right]$$

so that

$$\frac{1}{1 - Q(t)} = \exp\left(\frac{t - \gamma}{\eta}\right)^{\beta}$$

Therefore

$$\log\frac{1}{1 - Q(t)} = \left(\frac{t - \gamma}{\eta}\right)^{\beta}$$

and

$$\text{loglog}\frac{1}{1 - Q(t)} = \beta \log(t - \gamma) - \beta \log \eta$$

Which is $Y = mX + C$, the equation of a straight line.

If $(t - \gamma)$ is replaced by t' then:

$$Y = \text{loglog}\frac{1}{1 - Q(t)} \text{ and } X = \log t' \text{ and the slope } m = \beta.$$

If $Y = 0$

$$\text{loglog}\frac{1}{1 - Q(t)} = 0$$

then

$$\beta \log t' = \beta \log \eta$$

so that

$$t' = \eta$$

This occurs if

$$\text{loglog}\frac{1}{1 - Q(t)} = 0 \quad \text{so that} \quad \log\frac{1}{1 - Q(t)} = 1$$

i.e.

$$\frac{1}{1 - Q(t)} = e \quad \text{and} \quad Q(t) = 0.63$$

If a group of failures are distributed according to the Weibull function and it is initially assumed that $\gamma = 0$ then by plotting these failures against time on double logarithmic paper (failure percentage on loglog scale and time on log scale) a straight line should be obtained. The three Weibull parameters and hence the expression for reliability may then be obtained from measurements of the slope and intercept.

Figure 13.1 is loglog by log graph paper with suitable scales for cumulative

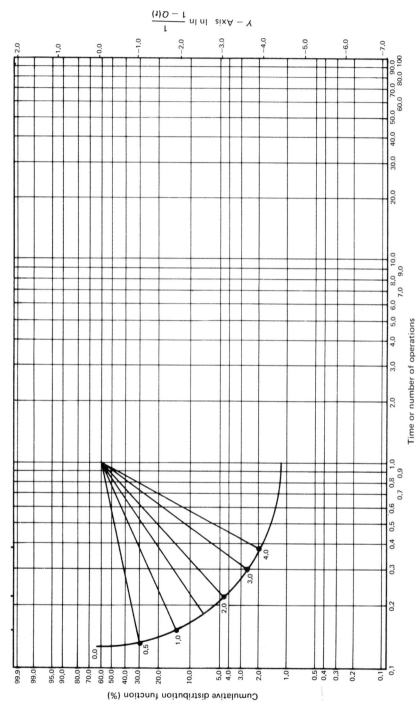

Figure 13.1   Graph paper for Weibull plot

percentage failure and time. Cumulative percentage failure is effectively the unreliability and is estimated by taking each failure in turn from median ranking tables of the appropriate sample size. It should be noted that the sample size, in this case, is the number of failures observed. A test yielding 10 failures from 25 items would require the first 10 terms of the median ranking table for sample size 25.

13.2 USING THE WEIBULL METHOD

Ten devices were put on test and permitted to fail without replacement. The time at which each device failed was noted and from the test information we require to determine:

(a) If there is a Weibull distribution which fits these data.
(b) If so, the values of γ, η and β.
(c) The probability of items surviving for specified lengths of time.
(d) If the failure rate is increasing, decreasing or constant.
(e) The MTBF.

The results are shown in table 13.1 against the median ranks for sample size 10. The ten points are plotted on Weibull paper as in figure 13.2 and a straight line is obtained.

Table 13.1

Cumulative failures, Q_t(%) median rank	6.7	16.2	25.9	35.6	45.2	54.8	64.5	74.1	83.8	93.3
Time, t (hours x 100)	1.7	3.5	5.0	6.4	8.0	9.6	11	13	18	22

The straight line tells us that the Weibull distribution is applicable and the parameters are determined as follows:

γ: It was shown in section 13.1 that if the data yields a straight line then
   $\gamma = 0$.
β: The slope yields the value of β. This is assisted by the construction on the
   graph paper. Values of β are marked on the arc for different slopes. Here
   $\beta = 1.5$.
η: In section 13.1 it was proved that $\eta = t$ for $Q(t) = 0.63$. Therefore
   $\eta = 11.1 \times 10^2 = 1110$ h.

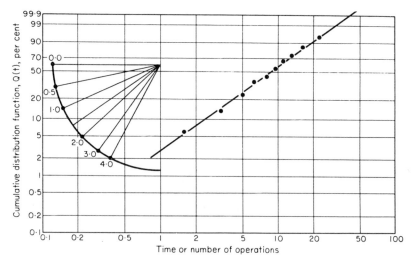

Figure 13.2   Results plotted on Weibull paper

The reliability expression is therefore:

$$R(t) = \exp \left[-\left(\frac{t}{1110} \right)^{1.5} \right]$$

The probability of survival to $t = 1000$ h is therefore:

$$R(1000) = e^{-0.855} = 42.5 \text{ per cent}$$

The test shows a wearout situation since:

For INCREASING FAILURE RATE	$\beta > 1$
For DECREASING FAILURE RATE	$\beta < 1$
For CONSTANT FIALURE RATE	$\beta = 1$

It now remains to evaluate the MTBF. This is, of course, the integral from zero to infinity of $R(t)$. Table 13.2 enables us to short cut this step.

Since $\beta = 1.5$ then MTBF/$\eta = 0.903$ and MTBF $= 0.903 \times 1110 = 1002$ h. Since median rank tables have been used the MTBF and reliability values calculated are at the 50 per cent confidence level. In the example time was recorded in hours but there is no reason why a more appropriate scale should not be used such as number of operations or cycles. The MTBF would then be quoted as Mean Number of Cycles between Failure.

For samples of other than 10 items a set of median ranking table is required. Since space does not permit a full set to be included the following

Table 13.2

β	$\dfrac{\text{MTBF}}{\eta}$	β	$\dfrac{\text{MTBF}}{\eta}$	β	$\dfrac{\text{MTBF}}{\eta}$	β	$\dfrac{\text{MTBF}}{\eta}$
0.0	∞	1.0	1.000	2.0	0.886	3.0	0.894
0.1	10!	1.1	0.965	2.1	0.886	3.1	0.894
0.2	5!	1.2	0.941	2.2	0.886	3.2	0.896
0.3	9.261	1.3	0.923	2.3	0.886	3.3	0.897
0.4	3.323	1.4	0.911	2.4	0.886	3.4	0.898
0.5	2.000	1.5	0.903	2.5	0.887	3.5	0.900
0.6	1.505	1.6	0.897	2.6	0.888	3.6	0.901
0.7	1.266	1.7	0.892	2.7	0.889	3.7	0.902
0.8	1.133	1.8	0.889	2.8	0.890	3.8	0.904
0.9	1.052	1.9	0.887	2.9	0.892	3.9	0.905
						4.0	0.906

approximation is given. For sample size N the rth rank is obtained from:

$$\frac{r - 0.3}{N + 0.4}$$

The dangers of attempting to construct a Weibull plot with too few points should be noted. A satisfactory result will not be obtained with less than at least six points. Tests yeilding 0, 1, 2 and even 3 failures do not enable any changing failure rate to be observed. In these cases constant failure rate must be assumed and the χ^2 test used. This is valid providing the information extracted is only applied to the same time range as the test.

13.3 MORE COMPLEX CASES OF THE WEIBULL DISTRIBUTION

Suppose that the data in our example had yielded a curve rather than a straight line. It is still possible that the Weibull distribution applies but with γ greater than zero. The approach is to select an assumed value for γ, usually the first value of t in the data, and replot the line against t', where $t' = t - \gamma$. The first point is now not available and the line will be constructed from one less point. Should the result be a straight line then the value of γ is as estimated and one proceeds as before to evaluate the other two parameters. MTBF is calculated as before plus the value of γ. If, on the other hand, another curve is generated then a further value of γ is tried until, by successive approximations, the correct value is found. This trial and error method of finding γ is not as time consuming as it might seem. It is seldom necessary to attempt more than four approximations of γ before either generating a straight line or confirming that the Weibull distribution will not fit the data. One possible reason for the Weibull distribution not applying could be the presence of more than one failure mechanism in the data. Two mechanisms are unlikely to follow the same distribution as each other and it is important to confine the analysis to one mechanism at a time.

So far a single sided analysis, at 50 per cent confidence has been described. It is possible to plot the 90 per cent confidence bands by use of the 5 and 95 per cent rank tables. First table 13.3 is constructed and the confidence bands plotted as follows.

Consider the point corresponding to the failure at 500 h. The two points A and B are marked on the straight line corresponding to 8.7 per cent and 51 per cent respectively. The median rank for this point was 25.9 per cent and vertical lines drawn from A and B to intersect the horizontal. These two points lay on the confidence bands. The other points are plotted in the same way and confidence bands produced as shown in figure 13.3. Looking at the curves the limits of $Q(t)$ at 1000 h are 30 per cent and 85 per cent. At 90 per cent confidence the reliability for 1000 h is therefore between 15 and 70 per cent.

Table 13.3

Time, t (hours x 100)	1.7	3.5	5.0	6.4	8.0	9.6	11	13	18	22
Median rank	6.7	16.2	25.9	35.6	45.2	54.8	64.5	74.1	83.8	93.3
5% rank	0.5	3.7	8.7	15	22	30	39	49	61	74
95% rank	26	39	51	61	70	78	85	91	96	99

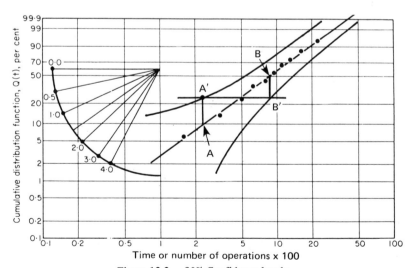

Figure 13.3   90% Confidence bands

13.4 SOME MAINTENANCE CONSIDERATIONS

If a particular component or equipment has a wearout characteristic and its
Weibull parameters have been obtained then it is possible to calculate the
probability of failure for any value of time. A time for preventive replacement
can be chosen against a given probability that the device will fail beforehand.

In some cases a population of variable failure rate devices, such as lamps, may
be regarded as failing at random. Imagine a building using a large number of
lamps where each is replaced as it fails so that, after several times the mean life
has elapsed, the population will be made up of lamps at different points in their
life. Figure 13.4 shows the superimposition of successive generations as a result of
which failures are occurring at random.

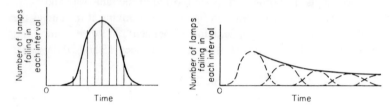

Figure 13.4   Failure distribution of a large population

EXERCISES

1. Components, as described in the example of section 13.2, are to be used in a
   system. It is required that these are preventively replaced such that there is
   only a 5 per cent probability of their failing beforehand. After how many
   hours should each item be replaced?

2. A sample of 10 items is allowed to fail and the time for each failure is as
   follows:

   4, 6, 8, 11, 12, 13, 15, 17, 20, 21 (thousand hours)

   Use the Weibull paper in chapter 13 to determine the reliability characteristic
   and the MTBF.

14 Demonstrating Maintainability

14.1 DEMONSTRATION RISKS

Where demonstration of a maintainability requirement is contractual it is essential that the test method, and the conditions under which it is to be carried out, are fully described. If this is not observed then disagreements are likely to arise during the demonstration. Both supplier and customer wish to achieve the specified Mean Time To Repair at minimum cost and yet a precise demonstration having acceptable risks to all parties is extremely expensive. A true assessment of maintainability can only be made at the end of the equipment life and anything less will represent a sample carrying the risks described in chapters 12 and 13.

Figure 14.1 shows a typical test plan for observing the Mean Time To Repair of a given item. Just as, in chapter 12, the curve shows the relationship of the probability of passing the test against the batch failure rate then figure 14.1 relates that probability to the actual MTTR.

For a MTTR of M_0 the probability of passing the test is 90 per cent and for a value of M_1 it falls to 10 per cent. In other words if M_0 and M_1 are within 2:1 of each other then the test has a good discrimination.

A fully documented procedure is essential and the only reference document available is US Military Standard 471A — Maintainability Verification/ Demonstration/Evaluation — 27 March 1973. This document may be used as the

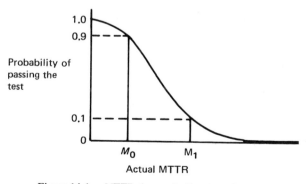

Figure 14.1   MTTR demonstration test plan

basis for a contractual agreement in which case both parties should carefully assess the risks involved. Statistical methods are usually dependent on assumptions concerning the practical world and it is important to establish their relevance to a particular test situation. In any maintainability demonstration test it is absolutely essential to fix the following:

Method of test demonstration task selection.
Tools and test equipment available.
Maintenance documentation.
Skill level and training of test subject.
Environment during test.
Preventive maintenance given to test system.

14.2 US MIL STANDARD 471A

This document replaces US MIL STANDARD 471 — 15 February 1966 — Maintainability Demonstration. It contains a number of sampling plans for demonstrating maintenance times for various assumptions of repair time distribution. A task sampling plan is also included and describes how the sample of simulated failures should be chosen. Test plans choose either the log normal assumption or make no assumption of distribution. The log normal distribution frequently applies to systems using consistent technologies such as computer and data systems, telecommunications equipment, control systems and consumer electronics but equipment with mixed technologies such as aircraft flight controls, microprocessor controlled mechanical equipment and so on are likely to exhibit bimodal distributions. This results from two repair time distributions (for two basic types of defect) being superimposed. Figure 14.2 illustrates this case.

The method of task sample selection involves stratified sampling. This involves dividing the equipment into functional units and, by ascribing failure rates to each unit, determining the relative frequency of each maintenance action. Taking into account the quantity of each unit the sample of tasks is spread according to the anticipated distribution of field failures. Random sampling is used to select specific tasks within each unit once the appropriate number of tasks has been assigned to each.

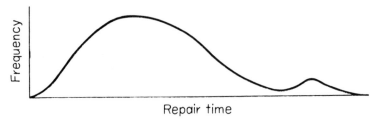

Figure 14.2   Distribution of repair times

The seven test plans are described as follows:

Test Method 1   The method tests for the mean repair time (MTTR). A minimum sample size of 30 is required and an equation is given for computing its value. Equations for the producer's and consumer's risks, α and β, and their associated repair times are also given. Two test plans are given. Plan A assumes a lognormal distribution of repair times whilst plan B is distribution free. That is to say it applies in all cases.

Test Method 2   The method tests for a percentile repair time. This means a repair time associated with a given probability of not being exceeded. For example a 90 percentile repair time of one hour means that 90 per cent of repairs are effected in one hour or less and that only 10 per cent exceed this value. This test assumes a lognormal distribution of repair times. Equations are given for calculating the sample size, the risks and their associated repair times.

Test Method 3   The method tests the percentile value of a specified repair time. It is distribution free and therefore applies in all cases. For a given repair time, values of sample size and pass criterion are calculated for given risks and stated pass and fail percentiles. For example if a median MTTR of 30 min is acceptable, and if 30 min as the 25th percentile (75 per cent of values are greater) is unacceptable, the test is established as follows. Producer's risk is the probability of rejection although 30 min is the median and Consumer's risk is the probability of acceptance although 30 min is only the 25th percentile. Let these both equal 10 per cent. Equations then give the value of sample size as 23 and the criterion as 14. Hence if more than 14 of the observed values exceed 30 min the test is failed.

Test Method 4   The method tests the median time. The median is the value, in any distribution, such that 50 per cent of values exceed it and 50 per cent do not. Only in the Normal distribution does the median equal the mean. A lognormal distribution is assumed in this test which has a fixed sample size of 20. The test involves comparing log MTTR in the test with log of the median value required in a given equation.

Test Method 5   The method tests the 'Chargeable Down Time per Flight'. This means the down time attributable to failures as opposed to passive maintenance activities, test induced failures, modifications, etc. It is distribution free with a minimum sample size of 50 and can be used, indirectly, to demonstrate availability.

Test Method 6   The method is applicable to aeronautical systems and tests the 'Manhour Rate'. This is defined as

$$\frac{\text{Total Chargeable Maintenance Manhours}}{\text{Total Demonstration Flight Hours}}$$

Actual data are used and no consumer or producer risks apply.

Test Method 7   This is similar to Test Method 6 and tests the manhour rate for simulated faults. There is a minimum sample size of 30.

Test Methods 1—4 are of a general nature whereas methods 5—7 have been developed with aeronautical systems in mind. In applying any test the risks must be carefully evaluated. There is a danger, however, of attaching an importance to results in proportion to the degree of care given to the calculations. It should therefore be emphasised that attention to the items listed in section 14.1 in order to ensure that they reflect the agreed maintenance environment is of equal if not greater importance.

14.3 DATA COLLECTION

It would be wasteful to regard the demonstration test as no more than a means of determining compliance with a specification. Each repair is a source of maintainability design evaluation and a potential input to the manual. Diagnostic instructions should not be regarded as static but be updated as failure information accrues. If the feedback is to be of use it is necessary to record each repair with the same detail as is called for in field reporting. The different repair elements of diagnosis, replacement, access, etc., should be separately listed together with details of tools and equipment used. Demonstration repairs are easier to control than field maintenance and should therefore be better documented.

In any maintainability, or reliability, test the details should be fully described in order to minimise the possibilities of disagreement. Both parties should understand fully the quantitative and qualitative risks involved.

15 Reliability Prediction

15.1 METHOD OF PREDICTION

Whilst it is component failure rate that is measured the reliability of complete equipment and systems is the ultimate concern of the designer and customer. Reliability prediction is the process of calculating the anticipated system reliability from assumed component failure rates. It provides a quantitative measure of how close a design comes to meeting the reliability objective and also permits comparisons between alternative design proposals. The simplest type of prediction involves little more than a parts count. Individual stress levels are not considered and an average failure rate for each component type is multiplied by the number involved. The overall total failure rate is used to calculate the system MTBF or reliability. It will be seen in section 15.3 that this simple addition of failure rates takes no account of redundancy and therefore gives a worst case prediction. It was mentioned in section 7.5 that failure rate data usually refers to random failures (flat portion of the bathtub). As a result 'parts count' reliability predictions involve constant failure rates and the summing of failure rates is permissible. This is not always the case and the exceptions to this procedure will be clearly explained in this chapter. As the design details become firmer more sophisticated predictions can be attempted taking account of failure modes, redundancy of parts and modules, stresses and environment and the quality and screening of components. Some examples of typical failure rate data are given in Appendix 3 and are expressed in terms of 10^{-9} per hour.

There are five steps involved in a reliability prediction.

15.1.1 Establish Failure Criteria

Define what constitutes a system failure since this will determine which failure modes at the component level actually cause a system failure. There may well be more than one type of system failure in which case a number of predictions giving different reliabilities will result. This step is essential if the stress and failure analysis and the subsequent predictions are to have any significance.

15.1.2 Establish a Reliability Block Diagram

It is necessary to represent the system as a number of functional blocks. These are interconnected according to the effect of each block failure on the total system.

Figure 15.1 shows a series diagram representing a system of *n* blocks such that the failure of any block prevents operation of the system. Alternatively figure 15.2 shows a situation where all blocks must fail in order for the system to fail. This is known as a parallel, or redundancy, case. A composite example is given in figure 15.3 which is a combination of series and parallel reliability. It represents a system which will fail if block A fails or if both block B and block C fail. The failure of B and C alone is insufficient to cause system failure.

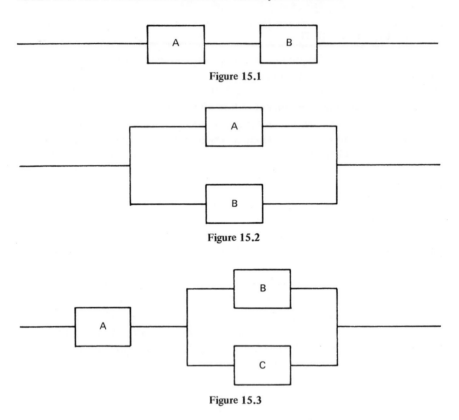

Figure 15.1

Figure 15.2

Figure 15.3

A number of general rules should be born in mind when defining the blocks.

(i)   Each block should represent the maximum number of components in order to simplify the diagram.
(ii)   The function of each block should be easily identified.
(iii) Blocks should be mutually independent in that failure in one should not effect the probability of failure in another.
(iv)   Blocks should not contain any significant redundancy otherwise the addition of failure rates would not be valid.
(v)   Each replaceable unit should be a whole number of blocks.

(vi)   Each block should contain one technology, that is electronic or electro-
mechanical.

(vii) There should be only one environment within a block.

15.1.3 Stress and Failure Analysis

The Failure Mode and Effect Analysis (FMEA) described in chapter 7 includes
the necessary stress and failure analysis required to generate block failure rates.
Given a constant failure rate and no internal redundancy each block will have a
failure rate predicted from the sum of the failure rates on the FMEA worksheet.

15.1.4 Calculation of System Reliability

Relating the block failure rates to the system reliability is a question of
mathematical modelling which is the subject of the rest of this chapter. In the
event that the system reliability prediction fails to meet the objective then
improved failure rate objectives must be assigned to each block by means of
allocation.

15.1.5 Reliability Allocation

The importance of reliability allocation was stressed in chapter 4 and an example
was calculated. The block failure rates established in section 15.1.3 are taken as
a measure of the complexity, and improved, suitably weighted, objectives are set.

15.2 PROBABILITY THEORY

The following basic probability rules are sufficient for an understanding of the
system modelling involved in reliability prediction.

15.2.1 The Multiplication Rule

If two or more events can occur simultaneously, and their individual probabilities
of occurring are known, then the probability of simultaneous events is the
product of the individual probabilities. The shaded area in figure 15.4 represents
the probability of events A and B both occurring.

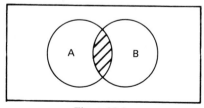

Figure 15.4

Hence the probability of A and B occurring is:

$$Pab = Pa \times Pb$$

Generally $$Pan = Pa \times Pb \ldots \ldots \times Pn$$

15.2.2 The Addition Rule

It is also required to calculate the probability of either event A OR event B OR BOTH occurring. This is the area of the two circles in figure 15.4. This probability is:

$$P(a \text{ or } b) = Pa + Pb - PaPb$$

being the sum of *Pa* and *Pb* less the area *PaPb* which is included twice. This becomes:

$$P(a \text{ or } b) = 1 - (1 - Pa)(1 - Pb)$$

Hence the probability of one or more of *n* events occurring is:

$$= 1 - (1 - Pa)(1 - Pb) \ldots \ldots (1 - Pn)$$

15.2.3 The Binomial Theorem

The above two rules are combined in the Binomial Theorem. Consider the following example involving a pack of 52 playing cards. A card is removed at random, its suit noted, and then replaced. A second card is then removed and its suit noted. The possible outcomes are:

Two hearts.
One heart and one other card.
Two other cards.

If *p* is the probability of drawing a heart then, from the multiplication rule, the outcomes of the experiment can be calculated as follows:

Probability of 2 hearts p^2
Probability of 1 heart $2pq$
Probability of 0 hearts q^2

Similar reasoning for an experiment involving 3 cards will yield:

Probability of 3 hearts p^3
Probability of 2 hearts $3p^2q$
Probability of 1 heart $3pq^2$
Probability of 0 hearts q^3

The above probabilities are the terms of the expressions $(p + q)^2$ and $(p + q)^3$. This leads to the general statement that if p is the probability of some random event, and if $q = 1 - p$, then the probabilities of 0, 1, 2, 3, ... outcomes of that event in n trials are given by the terms of the expansion:

$$(p + q)^n \text{ which equals}$$

$$p^n, \, np^{(n-1)}q, \, \frac{n(n-1)p^{(n-2)}q^2}{\text{factorial 2}}, \, \ldots \ldots, q^n$$

This is known as the binomial expansion.

15.2.4 Bayes Theorem

The marginal probability of an event is its simple probability. Consider a box of 7 cubes and 3 spheres in which case the marginal probability of drawing a cube is 0.7. To introduce the concept of a Conditional Probability assume that 4 of the cubes are black and 3 white and that, of the spheres, 2 are black and 1 is white as shown in figure 15.5.

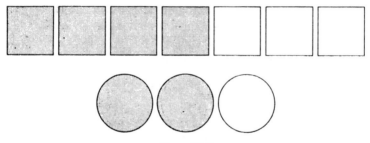

Figure 15.5

The probability of drawing a black article, given that it turns out to be a cube, is a conditional probability of 4/7 and ignores the possibility of drawing a sphere. Similarly the probability of drawing a black article, given that it turns out to be a sphere, is 2/3. On the other hand the probability of drawing a black sphere is a Joint Probability. It acknowledges the possibility of drawing cubes and spheres and is therefore 2/10.

Comparing joint and conditional probabilities, the conditional probability of drawing a black article given that it is a sphere is the joint probability of drawing a black sphere (2/10) divided by the probability of drawing any sphere (3/10). The result is hence $\frac{2}{3}$. Therefore:

$$P_{b/s} = \frac{P_{bs}}{P_s} \text{ given that:}$$

$P_{b/s}$ is the conditional probability of drawing a black article given that it is a sphere; P_s is the simple or marginal probability of drawing a sphere; P_{bs} is the joint probability of drawing an article which is both black and a sphere.

This is known as Baye's Theorem. It follows then that $P_{bs} = P_{b/s} \cdot P_s$ or $P_{s/b} \cdot P_b$ Consider now the probability of drawing a black sphere (P_{bs}) and the probability of drawing a white sphere (P_{ws}).

$$P_s = P_{bs} + P_{ws}$$

Therefore

$$P_s = P_{s/b} \cdot P_b + P_{s/w} \cdot P_w$$

and in general

$$P_x = P_{x/a} \cdot P_a + P_{x/b} \cdot P_b \ldots \ldots + P_{x/n} \cdot P_n$$

which is the form applicable to prediction formulae.

15.3 RELIABILITY OF SERIES SYSTEMS

The simple series system shown in figure 15.1 consists of two units, A and B, such that if either fails the system fails. The reliability of the system is the probability of unit A not failing and unit B not failing.

From the multiplication rule in section 15.2.1. then:

$$R_{ab} = R_a \cdot R_b \text{ and in general}$$
$$R_{an} = R_a \cdot R_b \ldots \ldots R_n$$

In the constant failure rate case where:

$$R_a = e^{-\lambda_a t}$$

Then $$R_n = \exp[-(\lambda_a + \lambda_b \ldots \ldots \lambda_n)t]$$

From which it can be seen that the system is also a constant failure rate unit whose reliability is of the form e^{-Kt} where K is the sum of the individual failure rates. Providing that the two assumptions of constant failure rate and series modelling apply then it is valid to talk of a system failure rate computed from the sum of the individual unit or component failure rates.

15.4 RELIABILITY OF SYSTEMS INVOLVING REDUNDANT UNITS

There are a number of ways in which redundancy can be applied as a means of improving reliability. These are summarised in figure 15.6.

These models are described in the following sections and calculate reliability and MTBF assuming that failed units are not subject to repair before enough units fail to constitute a system failure.

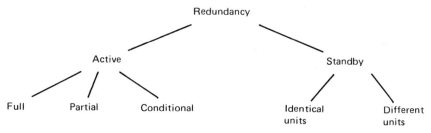

Figure 15.6   Redundancy

15.4.1 Full Active Redundancy

This is the situation where a number of units are functioning but system operation is maintained so long as one unit remains operating. Failure of all units is required for a system failure to occur. The addition rule in section 15.2.2 applies here since the reliability of such a system is the probability of non-failure of one or more of the units. Hence for n items:

$$R_n = 1 - (1 - R_a)(1 - R_b) \ldots \ldots (1 - R_n)$$

which for two identical items reduces to:

$$R_n = 1 - (1 - R_a)^2 = 2R_a - R_a{}^2$$

or for two different items to:

$$R_n = R_a + R_b - R_a R_b$$

If both items have constant failure rate, λ, the expression becomes:

$$R_n = 2e^{-\lambda t} - e^{-2\lambda t}$$

It is very important to note that, unlike the series case, this combination of constant failure rate units exhibits a reliability characteristic which is NOT of the form e^{-Kt}. In other words although constant failure rate units are involved the failure rate of the system is variable. The MTBF can therefore only be obtained from the integral of reliability. In chapter 3 we saw that

$$\text{MTBF} = \int_0^\infty R(t)\, dt$$

Hence

$$\text{MTBF} = \int_0^\infty (2e^{-\lambda t} - e^{-2\lambda t})$$

$$= 2/\lambda - 1/2\lambda$$

$$= 3/2\lambda$$

$$= 3\theta/2 \text{ where } \theta \text{ is the MBTF of a single unit.}$$

The danger now is to assume that the failure rate of the system is $2\lambda/3$. This is not true since the practice of inverting MTBF to obtain failure rate, and vice versa, is only valid for CONSTANT FAILURE RATE. Within the above working we substituted θ for $1/\lambda$ but in that case a unit was being considered for which constant λ applies.

Figure 15.7 compares reliability against time, and failure rate against time, for series and redundant cases. As can be seen the failure rate, initially zero, increases asymptotically. Reliability, in a redundant configuration, stays high at the beginning of a mission but eventually falls more sharply. The greater the number of redundant units the longer the period of higher reliability and the sharper the decline. These curves apply, in principle, to all redundant situations — only specific values change.

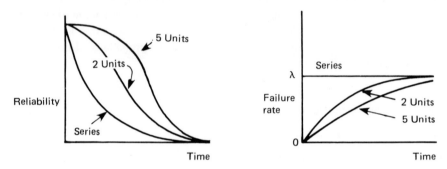

Figure 15.7   Effect of redundancy on reliability and failure rate

15.4.2 Partial Active Redundancy

Consider three identical units each with reliability R. Let $R + Q = 1$ so that Q is the unreliability (probability of failure in a given time). The Binomial expression $(R + Q)^3$ yields the following terms:

$$R^3, 3R^2Q, 3RQ^2, Q^3 \text{ which are}$$
$$R^3, 3R^2(1-R), 3R(1-R)^2, (1-R)^3$$

This conveniently describes the probabilities of

$$0 , 1 \quad , \quad 2 \quad , \quad 3 \quad \text{failures of a single unit.}$$

In section 15.4.1 the reliability for full redundancy was seen to be:

$$1 - (1-R)^3$$

This is consistent with the above since it can be seen to be 1 minus the last term. Since the sum of the terms is unity reliability is therefore the sum of the first three terms which, being the probability of 0, 1 or 2 failures, is the reliability of a fully redundant system.

In many cases of redundancy, however, the number of units permitted to fail before system failure occurs is less than in full redundancy. In the example of three units full redundancy requires only one to function whereas partial redundancy would exist if 2 units were required with only one allowed to fail. Once again the reliability can be obtained from the binomial expression since it is the probability of 0 or 1 failures which is given by the sum of the first 2 terms. Hence:

$$R_{system} = R^3 + 3R^2(1 - R)$$
$$= 3R^2 - 2R^3$$

In general if r items may fail out of n then the reliability is given as the sum of the first $r + 1$ terms of the binomial expansion $(R + Q)^n$. Therefore

$$R = R^n + nR^{n-1}(1 - R) + \frac{n(n - 1)R^{n-2}(1 - R)^2}{\text{Factorial } 2} + \cdots$$

$$\cdots + \frac{n(n - 1) \ldots (n - r + 1)R^{n-r}(1 - R)^r}{\text{Factorial } r}$$

15.4.3 Conditional Active Redundancy

This is best considered by an example. Consider the configuration in figure 15.8.

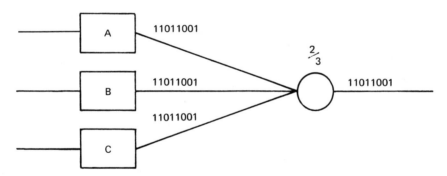

Figure 15.8

Three identical digital processing units (A, B and C) have reliability R. They are triplicated to provide redundancy in the event of failure and their identical outputs are fed to a 2 out of 3 majority voting gate. If two identical signals are received by the gate they are reproduced at the output. Assume that the voting gate is sufficiently more reliable than the units so that its probability of failure can be disregarded. Assume also that the individual units can fail either to an open circuit or a short circuit output. Random data bit errors are not included in the definition of system failure for the purpose of this example. The question arises as to whether the system has:

  (a) Partial Redundancy   1 unit may fail but no more.
or (b) Full Redundancy   2 units may fail.

The answer is conditional on the mode of failure. If two units fail in a like mode (both outputs logic 1 or logic 0) then the output of the voting gate will be held at the same value and the system will have failed. If, on the other hand, they fail in unlike modes then the remaining unit will produce a correct output from the gate since it always sees an identical binary bit from one of the other units. This conditional situation requires Bayes theorem for a calculation of reliability. The equation in section 15.2.4. becomes:

$$R_{system} = R_{given\ A} \cdot P_A + R_{given\ B} \cdot P_B \cdots \cdots + R_{given\ N} \cdot P_N$$

$$\text{where } A \text{ to } N \text{ are mutually exclusive and } \sum_{i=A}^{i=N} P_i = 1$$

In this case the solution is:

$$R_{system} = R_{\substack{\text{system given that in the event of failure} \\ \text{2 units fail alike}}} \times P_{\text{failing alike}}$$
$$+$$
$$+ R_{\substack{\text{system given that in the event of failure} \\ \text{2 units fail unalike}}} \times P_{\text{failing unalike}}$$

Therefore:

$$R_s = [R^3 + 3R^2(1-R)] \cdot P_A + [1 - (1-R)^3] \cdot P_B$$

Since if two units fail alike there is partial redundancy and if two units fail unalike there is full redundancy. Assume that the probability of both failure modes is the same and that $P_A = P_B = 0.5$. The system reliability is therefore:

$$R_s = \frac{R^3 + 3R^2 - 3R^3 + 1 - 1 + 3R - 3R^2 + R^3}{2} = \frac{3R - R^3}{2}$$

15.4.4 Standby Redundancy

So far only active redundancy has been considered where every unit is operating and the system can function despite the loss of one or more units. Standby redundancy involves additional units which are only activated when the operating unit fails. A greater improvement, per added unit, is anticipated than with active redundancy since the standby units operate for less time. Figure 15.9 shows n identical units with item 1 active. Should a failure be detected then item 2 will be switched in its place. Initially the following assumptions are made:

1. The means of sensing that a failure has occurred and for switching from the defective to the standby unit is assumed to be failure free.

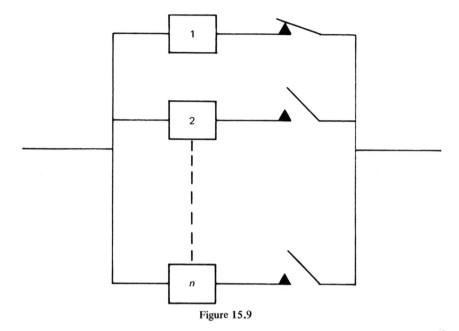

Figure 15.9

2. The standby unit(s) are assumed to have identical, constant failure rates to the main unit.
3. The standby units are assumed not to fail whilst in the idle state.
4. As with the earlier calculation of active redundancy, defective units are assumed to remain so. No repair is effected until the system has failed.
   Calculations involving redundancy and repair are covered in later sections.
The reliability is then given by the first n terms of the Poisson expression:

$$R_{system} = R(t) = e^{-\lambda t}\left(1 + \lambda t + \frac{\lambda^2 t^2}{2!} \cdots \frac{\lambda^{(n-1)} t^{(n-1)}}{(n-1)!}\right)$$

which reduces, for two units to:

$$R_{system} = e^{-\lambda t}(1 + \lambda t)$$

Figure 15.10 shows the more general case of two units with some of the above assumptions removed.

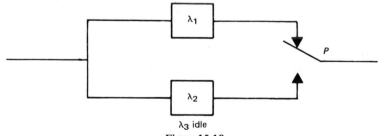

λ_3 idle

Figure 15.10

λ_1 is the constant failure rate of the main unit.
λ_2 is the constant failure rate of the standby unit when in use.
λ_3 is the constant failure rate of the standby unit in the idle state.
P is the one shot probability of the switch performing when required.

The reliability is given by:

$$R_{\text{system}} = e^{-\lambda_1 t} + \frac{P\lambda_1}{\lambda_2 - \lambda_1 - \lambda_3} \left(e^{-(\lambda_1 + \lambda_3)t} - e^{-\lambda_2 t} \right)$$

It remains only to consider the following failure possibilities: Let λ_4, λ_5 and λ_6 be the failure rates associated with the sums of the following failure modes:

For λ_4 – Dormant failures which inhibit failure sensing or changeover.
For λ_5 – Failures causing the incorrect switching back to the failed unit.
For λ_6 – False sensing of non-existent failure.

If we think about each of these in turn it will be seen that, from the point of view of the above model:

λ_4 is part of λ_3
λ_5 is part of λ_2
λ_6 is part of λ_1

In the stress and failure analysis they should therefore be added in to the appropriate category.

15.4.5 In General

Incremental Improvement   As was seen in figure 15.7 the improvement given by redundancy is not spread evenly along the time axis. Since the MTBF is an overall parameter obtained by integrating reliability from zero to infinity it is actually the area under the curve of reliability against time. For short missions (less than one MTBF in duration) the actual improvement in reliability is greater than would be indicated by comparing MTBFs. For this reason the length of mission should be considered when calculating reliability and deciding upon redundancy.

As we saw in section 15.4.1 the effect of duplicating a unit by active redundancy is only to improve the MTBF by 50 per cent. This improvement falls off as the number of redundant units increases as is illustrated in Figure 15.11. The effect is similar for other redundant configurations such as conditional and standby. Beyond a few units the improvement may even be offset by the unreliability introduced as a result of additional switching.

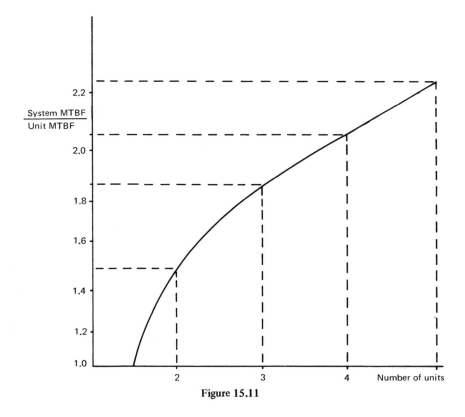

Figure 15.11

Comparisons of Redundancy   Figure 15.12 shows two approaches to system configuration using 4 units in active redundancy. (i) protects against short circuit failures whereas (ii) protects against short and open circuit conditions. As can be seen from Fig. 15.13 (ii) has the higher reliability but it is harder to design. The reader may care to calculate the MTBF of (i) and will find that it is less than for a single unit and, as can be seen from the diagram, the area under the reliability curve (MTBF) is less. It is of value only for conditions where short circuit failure is likely.

Figure 15.14 gives a comparison between a given number of units in both standby and active redundancy. For the simple model involving perfect switching

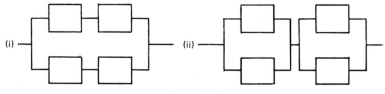

Figure 15.12

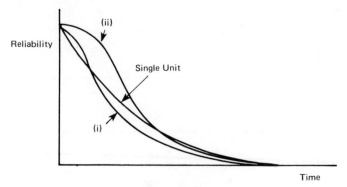

Figure 15.13

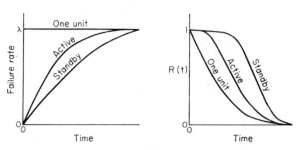

Figure 15.14

the standby configuration has the higher reliability although in practice the associated hardware for sensing and switching can erode the advantage. On the other hand it is not always easy to achieve active redundancy with true independence between units. In other words, the failure of one unit may cause or at least hasten the failure of another. This effect will be explained in the next sub-section.

Load Sharing   The following situation can be deceptive since, at first sight, it appears as active redundancy. Figure 15.15 shows two capacitors connected in series. Given that both must fail short circuit in order for the system to fail we require a model for the system. It is NOT two units in active redundant configuration because if the first capacitor should fail (short circuit) then the

Figure 15.15

voltage applied to the remaining one will be doubled and its failure rate greatly increased. This situation is known as load sharing and is mathematically identical to a standby arrangement. Figure 15.16 shows two units in standby configuration. The switchover is assumed to be perfect (which is appropriate) and the standby unit has an idle failure rate equal to zero with a different (larger) failure rate after switchover. The main unit has a failure rate of twice the single capacitor.

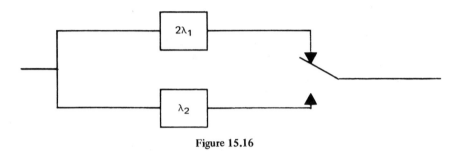

Figure 15.16

15.5 SYSTEMS WITH REDUNDANT UNITS AND PERIODIC REPAIR

It is often convenient, with unattended equipment, not to repair redundant items immediately they fail but to visit at regular intervals for the purpose of repairing or replacing failed redundant units. A system failure occurs when the redundancy is insufficient to sustain operation between visits. This is not as effective as immediate repair but costs considerably less in maintenance effort. If a system with redundant units and reliability $R(t)$ is visited every T h then:

$$\text{System MTBF} = \frac{\displaystyle\int_0^T R(t)\,dt}{1 - R(T)}$$

where $R(t)$ is the reliability without repair as calculated in the foregoing sections and $R(T)$ is the probability of survival for T h.

As an example take two constant failure rate units in active redundancy, visited every T h.

$$\text{System MTBF} = \frac{\displaystyle\int_0^T (2e^{-\lambda t} - e^{-2\lambda t})\,dt}{1 - 2e^{-\lambda T} + e^{-2\lambda T}}$$

$$= \frac{\dfrac{e^{-2\lambda T}}{2\lambda} - \dfrac{2e^{-\lambda T}}{\lambda} + \dfrac{2}{\lambda} - \dfrac{1}{2\lambda}}{1 - 2e^{-\lambda T} + e^{-2\lambda T}}$$

In a redundant configuration failure rate increases with time as was shown in figure 15.7. During the visits the system is restored to the fully operating condition at the beginning of each period and the failure rate assumed to return to zero. The graph of failure rate against time is therefore as shown in figure 15.17 from which it can be seen that the system may be assumed to have constant failure rate for times well in excess of T. This equivalent failure rate may be computed from 1/MTBF.

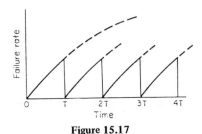

Figure 15.17

Figure 15.18 shows the system MTBF against T, the time between visits. As can be seen from the steep portion of the curve a small decrease in time between inspections can yield a spectacular increase in MTBF.

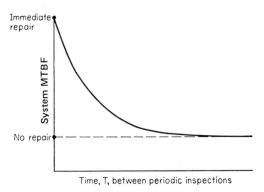

Figure 15.18

15.6 ATTENDED SYSTEMS WITH REDUNDANT UNITS

15.6.1 The effect of MTTR on System MTBF

By repairing the failed units in a redundant system a spectacular increase in reliability can be achieved, since the system is 'vulnerable' only during the down time of the failed redundant unit. There is therefore a relationship between reliability and down time and hence with MTTR. In the following equations the repair rate (μ) is the reciprocal of down time. The necessary additions to the repair time must be borne in mind when deriving down time. Before describing the method of calculation a number of expressions for system MTBF are given in terms of μ, and the constant failure rate, λ, of the units. Note that in most cases the simplified expression is given for the case of μ being much greater than λ. This is usually applicable in practice by at least an order of magnitude. Not only are the equations much simplified but the system may then be regarded as having constant failure rate.

(i) Active redundancy – two identical units.

$$= \frac{3\lambda + \mu}{2\lambda^2} \cong \frac{\mu}{2\lambda^2} \quad \text{if} \quad \mu \gg \lambda$$

(ii) Full active redundancy – n identical units.

$$\theta = \frac{\mu^{(n-1)}}{n!\lambda^n} \quad \text{if} \quad \mu \gg \lambda$$

(iii) Standby redundancy – two identical units.

$$\theta = \frac{2\lambda + \mu}{\lambda^2} \cong \frac{\mu}{\lambda^2} \quad \text{if} \quad \mu \gg \lambda$$

(iv) Standby redundancy – n identical units.

$$\theta = \frac{\mu^{(n-1)}}{\lambda^n} \quad \text{if} \quad \mu \gg \lambda$$

(v) Active redundancy – two different units

$$\theta = \frac{(\lambda_a + \mu_b)(\lambda_b + \mu_a) + \lambda_a(\lambda_a + \mu_b) + \lambda_b(\lambda_b + \mu_a)}{\lambda_a\lambda_b(\lambda_a + \lambda_b + \mu_a + \mu_b)}$$

(vi) Partial active redundancy

Total number of units				
1	$\dfrac{1}{\lambda}$	MTBF Table		
2	$\dfrac{3\lambda + \mu}{2\lambda^2}$	$\dfrac{1}{2\lambda}$		
3	$\dfrac{11\lambda^2 + 7\lambda\mu + 2\mu^2}{6\lambda^3}$	$\dfrac{5\lambda + \mu}{6\lambda^2}$	$\dfrac{1}{3\lambda}$	
4	$\dfrac{25\lambda^3 + 23\lambda^2\mu + 13\lambda\mu^2 + 3\mu^3}{12\lambda^4}$	$\dfrac{13\lambda^2 + 5\lambda\mu + \mu^2}{12\lambda^3}$	$\dfrac{7\lambda + \mu}{12\lambda^2}$	$\dfrac{1}{4\lambda}$
	1	2	3	4

Number of units required to operate

15.6.2 Method of Calculation

The probability of system failure will decrease with the MTTR. The calculation of the probability of survival, and hence MTBF, is as follows.

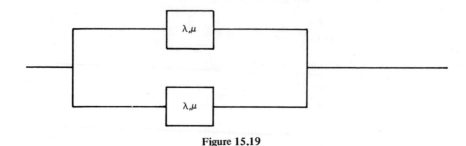

Figure 15.19

Figure 15.19 shows a system with two identical units, each having failure rate λ and repair rate μ. The system can therefore be in three possible states.

State (0)   Both units operating.
State (1)   One unit operating, the other having failed.
State (2)   Both units failed.

Let $P_i(t)$ be the probability that the system is in state (i) at time t and assume that the initial state is (0).

Therefore $P_0(0) = 1$ and $P_1(0) = P_2(0) = 0$

Therefore $P_0(t) + P_1(t) + P_2(t) = 1$

We shall now calculate the probability of the system being in each of the three states at time $t + \Delta t$.

The system will be in state (0) at time $t + \Delta t$ if:

(a) The system was in state (0) at time t and no failure occurred in either unit during the interval Δt, or,
(b) The system was in state (1) at time t, no further failure occurred during Δt, and the failed unit was repaired during Δt.

The probability of only one failure occurring in one unit during that interval is simply $\lambda \Delta t$ (valid if Δt is small, which it is). Consequently $(1 - \lambda \Delta t)$ is the probability that no failure will occur in one unit during the interval. The probability that both units will be failure free during the interval is, therefore,

$$(1 - \lambda \Delta t)(1 - \lambda \Delta t) \approx 1 - 2\lambda \Delta t$$

The probability that one failed unit will be repaired within Δt is $\mu \Delta t$, provided that Δt is very small. This leads to the equation:

$$P_0(t + \Delta t) = [P_0(t) \times (1 - 2\lambda\Delta t)] + [P_1(t) \times (1 - \lambda\Delta t) \times \mu\Delta t]$$

Similarly for states 1 and 2:

$$P_1(t + \Delta t) = [P_0(t) \times 2\lambda\Delta t] + [P_1(t)$$
$$\times (1 - \lambda\Delta t) \times (1 - \mu\Delta t)]$$
$$P_2(t + \Delta t) = [P_1(t) \times \lambda\Delta t] + P_2(t)$$

Now the limit as $\Delta t \to 0$ of $\dfrac{P_i(t + \Delta t) - P_i(t)}{\Delta t}$ is $\dot{P}_i(t)$ and so the above yeild:

$$\dot{P}_0(t) = -2\lambda P_0(t) + \mu P_1(t)$$
$$\dot{P}_1(t) = 2\lambda P_0(t) - (\lambda + \mu)P_1(t)$$
$$\dot{P}_2(t) = P_1(t)\lambda$$

In matrix notation this becomes:

$$
\begin{vmatrix} \dot{P}_0 \\ \dot{P}_1 \\ \dot{P}_2 \end{vmatrix}
=
\begin{vmatrix} -2\lambda & \mu & 0 \\ 2\lambda & -(\lambda + \mu) & 0 \\ 0 & \lambda & 0 \end{vmatrix}
\begin{vmatrix} P_0 \\ P_1 \\ P_2 \end{vmatrix}
$$

The elements of this matrix can also be obtained by means of a Transition Diagram. Since only one event can take place during a small interval, Δt, the transitions between states involving only one repair or one failure are considered. Consequently the transitions (with transition rates) are:

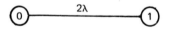

by failure of either unit

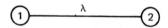

by failure of the remaining active unit,

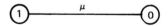

by repair of the failed unit of state 1.

The transition diagram is:

Finally closed loops are drawn at states 0 and 1 to account for the probability of not changing state. The rates are easily calculated as minus the algebraic sum of the rates associated with the the lines leaving that state. Hence:

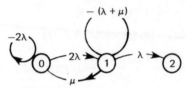

A (3×3) matrix, $(a_{i,j})$, can now be constructed, where: $i = 1, 2, 3; j = 1, 2, 3;$ $a_{i,j}$ is the character on the flow line pointing from state j to state i. If no flow line exists the corresponding matrix element is zero. We therefore find the same matrix as before.

The MTBF is defined as

$$\theta_s = \int_0^\infty R(t)\, dt$$

$$= \int_0^\infty [P_0(t) + P_1(t)]\, dt$$

$$= \int_0^\infty P_0(t)\, dt + \int_0^\infty P_1(t)\, dt$$

$$= T_0 + T_1$$

The values of T_0 and T_1 can be found by solving the following:

$$\int_0^\infty \begin{vmatrix} \dot{P}_0(t) \\ \dot{P}_1(t) \\ \dot{P}_2(t) \end{vmatrix} dt = \int_0^\infty \begin{vmatrix} -2\lambda & \mu & 0 \\ 2\lambda & -(\lambda + \mu) & 0 \\ 0 & \lambda & 0 \end{vmatrix} \begin{vmatrix} P_0 \\ P_1 \\ P_2 \end{vmatrix} dt$$

Since the (3×3) matrix is constant we may write

$$\int_0^\infty \begin{vmatrix} \dot{P}_0(t) \\ \dot{P}_1(t) \\ \dot{P}_2(t) \end{vmatrix} dt = \begin{vmatrix} -2\lambda & \mu & 0 \\ 2\lambda & -(\lambda + \mu) & 0 \\ 0 & \lambda & 0 \end{vmatrix} \int_0^\infty \begin{vmatrix} P_0 \\ P_1 \\ P_2 \end{vmatrix} dt$$

or

$$\begin{vmatrix} \int_0^\infty \dot{P}_0(t)\,dt \\[2mm] \int_0^\infty \dot{P}_1(t)\,dt \\[2mm] \int_0^\infty \dot{P}_2(t)\,dt \end{vmatrix} = \begin{vmatrix} -2\lambda & \mu & 0 \\ 2\lambda & -(\lambda+\mu) & 0 \\ 0 & \lambda & 0 \end{vmatrix} \begin{vmatrix} \int_0^\infty P_0(t)\,dt \\[2mm] \int_0^\infty P_1(t)\,dt \\[2mm] \int_0^\infty P_2(t)\,dt \end{vmatrix}$$

or

$$\begin{vmatrix} P_0(\infty) - P_0(0) \\ P_1(\infty) - P_1(0) \\ P_2(\infty) - P_2(0) \end{vmatrix} = \begin{vmatrix} -2\lambda & \mu & 0 \\ 2\lambda & -(\lambda+\mu) & 0 \\ 0 & \lambda & 0 \end{vmatrix} \begin{vmatrix} T_0 \\ T_1 \\ T_2 \end{vmatrix}$$

Taking account of $P_0(0) = 1; P_1(0) = P_2(0) = 0$

$$P_0(\infty) = P_1(\infty) = 0; P_2(\infty) = 1$$

we may reduce the equation to

$$\begin{vmatrix} -1 \\ 0 \\ 1 \end{vmatrix} = \begin{vmatrix} -2\lambda & \mu & 0 \\ 2\lambda & -(\lambda+\mu) & 0 \\ 0 & \lambda & 0 \end{vmatrix} \begin{vmatrix} T_0 \\ T_1 \\ T_2 \end{vmatrix}$$

or

$$-1 = -2\lambda T_0 + \mu T_1$$
$$0 = 2\lambda T_0 - (\lambda + \mu)T_1$$
$$1 = \lambda T_1$$

Solving this set of equations

$$T_0 = \frac{\lambda+\mu}{2\lambda^2} \quad \text{and} \quad T_1 = \frac{1}{\lambda}$$

so that

$$\theta_s = T_0 + T_1 = \frac{1}{\lambda} + \frac{\lambda+\mu}{2\lambda^2} = \frac{3\lambda+\mu}{2\lambda^2}$$

that is

$$\theta_s = \frac{3\lambda+\mu}{2\lambda^2}$$

15.7 PREDICTION IN PERSPECTIVE

15.7.1 The Numbers Game

It must be stressed that prediction is a design tool and not a precise measure of reliability. The main value of a prediction is in showing the relative reliabilities of modules so that allocations can be made. Whatever the accuracy of the exercise, if one module is shown to have double the MTBF of another then, when calculating values for modules in order to achieve the desired system MTBF, the values allocated to the modules should be in the same ratio. Prediction also permits a reliability comparison between different design solutions. Again, the comparison is likely to be accurate even if the absolute values are not. Concerning the accuracy of the actual predicted value this will depend on:

(a) Relevance of the failure rate data and the chosen environmental multiplication factors.
(b) Accuracy of the Mathematical Model.

The greater the number of different component types involved the more likely that individual over and underestimates will cancel each other out.

15.7.2 Computer Aids

Computer-aided reliability prediction is now available and is justified for any calculation which takes longer than to input the necessary system data to the computer. In practice any prediction which is more complex than can be achieved by an hour's use of the simple equations already given will justify a computer solution. Failure rates can be inputted with the system information or, in some cases, programs make use of failure rate data banks already stored. These are updated on a regular basis with fresh field data. The necessary statistical treatment of the data is carried out by the computer program.

EXERCISES

1. Calculate the MTBF of the system shown in the following block diagram.

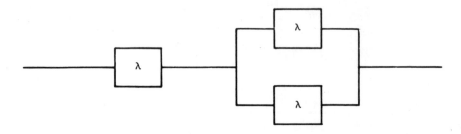

2. The following block diagram shows a system whereby unit B may operate with units D or E but where unit A may only operate with unit D, or C with E. Derive the reliability expression.

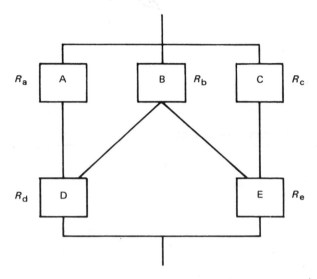

3. Calculate the MTBF of the system in Exercise 1 if it is visited at intervals, T, for repair of failed redundant units.

16 Prediction of Repair Times

Maintainability prediction is by no means as well developed as reliability prediction. The only fully developed systems are described in US MIL HDBK 472 which is dated 1966. In 1973 this document was withdrawn and replaced by US MIL STD 471A which contains no prediction techniques. The methods described in US MIL HDBK 472, although applicable to a range of equipment developed at that time, have much to recommend them and are still worthy of attention. Unfortunately the quantity of data required to develop these methods of prediction is so great that with increasing costs and shorter design lives the author fears that such exercises may not be repeated. On the other hand calculations requiring the statistical analysis of large quantities of data lend themselves to computer methods and the rapid increase of these facilities makes such a calculation feasible if the necessary repair time data for a very large sample of repairs (say 10 000) were available.

Any realistic maintainability prediction procedure must meet the following essential requirements:

(i) The prediction must be fully documented and described and must be subject to recorded modification as a result of experience.
(ii) All assumptions must be recorded and their validity checked where possible.
(iii) The prediction must be carried out by engineers who are not part of the design group and therefore not biased by the objectives.

Prediction, valuable as it is, should be replaced by demonstration as soon as possible in the programme. Maintainability is related to reliability in that the frequency of each repair action is determined by failure rates. Maintainability prediction therefore requires a knowledge of failure rates in order to select the appropriate, weighted, sample of tasks. The prediction results can therefore be no more reliable than the accuracy of the failure rate data.

Maintainability prediction is only applicable to the active elements of repair time since it is those which are influenced by the design. There are two approaches to the prediction task. The first is a work study method which analyses each task in the sample by breaking it into definable work elements. This requires an extensive data bank of average times for a wide range of tasks on the equipment type in question. The second approach is empirical and involves rating a number of maintainability factors against a checklist. The resulting 'scores' are converted to an MTTR by means of a nomograph which was obtained by regression analysis of the data.

The procedures in US MIL HANDBOOK 472 are over 10 years old and it is unlikely that the data bases are relevant to modern equipment. In the absence of alternative methods, however, procedure 3 is recommended because the prediction will give an indication of the repair time and also since the checklist approach to the method focuses attention on the practical features affecting repair time. Procedures 2 and 3 of US MIL 472 are described.

16.2 US MIL HANDBOOK 472 – Procedure 2

This procedure was developed by ITT for the United States Navy: it was intended for shipboard and shore electronic equipment. It was intended for use during the final stage of design and is a detailed work study process capable of giving accurate predictions provided that adequate funds and data were available. It is unlikely that it could be applied to equipment which breaks new ground or where funding of maintainability analysis is less than lavish. The data in the handbook, in view of their age, are unlikely to be suitable for current designs.

The method represents a comprehensive approach and is as near to an ideal prediction as possible. It deals with the relationship between the ERT (Equipment Repair Time), which is defined as the median of the individual repair times, and the MTTR (Mean Time To Repair) for different distributions of repair times. The following information is essential for the use of procedure 2.

(i) The structure and breakdown into groups, units, assemblies and components to the level of the LRA.
(ii) The detailed procedure for diagnosing a fault.
(iii) Repair methods to be used.
(iv) Sufficient information for a full reliability prediction based on a stress analysis.
(v) Mechanical details of the mountings and assemblies within the equipment.

The procedure breaks the active repair time into:

(i) Localisation – Determination of fault location without test gear.
(ii) Isolation – Location of the fault using external test equipment.
(iii) Access – Access to the part or LRA which has to be removed.
(iv) Interchange – Replacement of LRA.
(v) Reassembly – As necessary.
(vi) Alignment – As necessary.
(vii) Checkout – To confirm system performance meets requirements.

Worksheets A and B are shown in figures 16.1 and 16.2 and are used for the

Figure 16.1

Worksheet B

Contractor_____   Date_____

Contract No._____   Sheet_____of_____

1 Worksheet A Sheet Number	2 $\Sigma\lambda$ Column C Total	3 $\Sigma\lambda M_c$ Column M Total	1 Worksheet A Sheet Number	2 $\Sigma\lambda$ Column C Total	3 $\Sigma\lambda M_c$ Column M Total
Subtotals			Subtotals		

Product failure rate, $\Sigma\lambda$ = Total of column 2 subtotals_____
Total repair time per 10^6 hours, $\Sigma\lambda M_c$ = Total of column 3 subtotals_____

Figure 16.2

calculation of active corrective maintenance times. In column A of sheet A every LRA is entered. If repair is to component level then the list would include terminals, wiring and cables. Where the failure mode will effect the repair time separate entries (and times) must be made for each mode and the appropriate partial failure rates entered in column C. Failure mode is likely to effect localisation (diagnosis time) in column D due to the different symptoms produced. Columns D to K are completed in accordance with the data available and column L gives the total maintenance time for the condition being considered in that line. Column M contains the product of the failure rate for that mode and the time to repair. On each sheet the sum of the failure rates in column C and the sum of the products in column M are recorded.

Worksheet B is a summary sheet and is used to obtain $\Sigma\lambda$ and $\Sigma(\lambda M_c)$ for the whole equipment. The mean time to repair is predicted by the formula:

$$M_c = \frac{\Sigma(\lambda M_c)}{\Sigma\lambda}$$

Preventive maintenance times are predicted in a similar manner.

16.3 US MIL HANDBOOK 472 – PROCEDURE 3

Procedure 3 was developed by RCA for the US Air Force and was intended for ground systems. It requires a fair knowledge of the design detail and maintenance procedures for the system being analysed. The method is based on the principle of predicting a sample of the maintenance tasks. It is entirely empirical since it was developed to agree with known repair times for specific systems including search radar, data processors and a digital data transmitter with r.f. elements. The sample of repair tasks is selected on the basis of failure rates and it is assumed that the time to diagnose and correct a failure of a given component is the same as for any other of that component type. This is not always true as field data can show.

Where repair of the system is achieved by replacement of sizeable modules (that is a large LRA) the sample is based on the failure rate of these high level units.

The predicted repair time for each sample task is arrived at by considering a checklist of maintainability features and by scoring points for each feature. The score for each feature increases with the degree of conformity with a stated 'ideal'. The items in the checklist are grouped under three headings. These are Design, Maintenance Support and Personnel Requirements. The points scored under each heading are appropriately weighted and related to the predicted repair time by means of a regression equation which is presented in the form of an easily used nomograph.

Figure 16.3 shows the score sheet for use with the checklist and figure 16.4 presents the regression equation nomograph. Looking at the checklist it will be noticed that additional weight is given to some features of design or maintenance support by the fact that more than one score is influenced by a particular feature.

The checklist is reproduced, in part, in the following section but the reader wishing to carry out a prediction will need a copy of US MIL HANDBOOK 472 for the full list. The application of the checklist to typical tasks is, in the author's opinion, justified as an aid to maintainability design even if repair time prediction is not specifically required.

Equip._____ Unit/Part_____ Task No._____

Ass'y_____ By_____ Date_____

Primary function failed unit/part_____

_____

Mode of failure_____

Malfunction symptoms_____

_____

Maintenance Analysis

Maintenance Steps	Scoring Comments

Checklist Scores

	1	2	3	4	5	6	7	8	9	10	11	12	13	14	15	Total
A																
B								▨	▨	▨	▨	▨	▨	▨		
C											▨	▨	▨	▨	▨	

Predicted downtime   ____ Min.

Figure 16.3

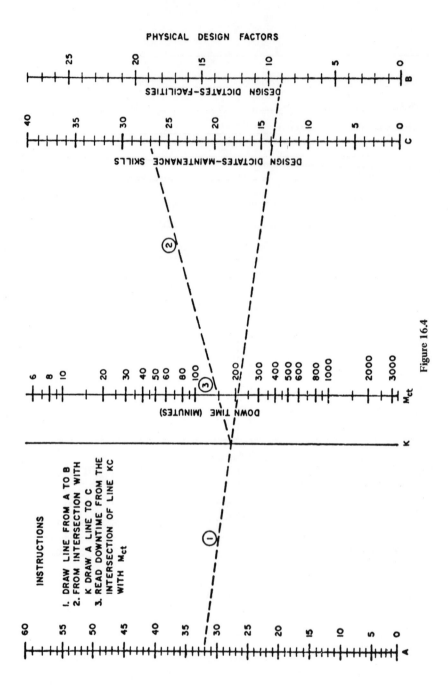

Figure 16.4

16.4 CHECKLIST – MIL 472 PROCEDURE 3

The headings of each of the checklists are as follows:

Checklist A

 1. Access (external)
 2. Latches and fasteners (external)
 3. Latches and fasteners (internal)
 4. Access (internal)
 5. Packaging
 6. Units/parts (failed)
 7. Visual displays
 8. Fault and operation indicators
 9. Test points availability
10. Test points identification
11. Labelling
12. Adjustments
13. Testing in circuit
14. Protective devices
15. Safety–personnel

Checklist B

 1. External test equipment
 2. Connectors
 3. Jigs and fixtures
 4. Visual contact
 5. Assistance operations
 6. Assistance technical
 7. Assistance supervisory

Checklist C

 1. Arm-leg-back strength
 2. Endurance and energy
 3. Eye-hand
 4. Visual
 5. Logic
 6. Memory
 7. Planning
 8. Precision
 9. Patience
10. Initiative

Three times from each of checklists A and B and the scoring criteria for all of checklist C are reproduced as follows. The reader may care to compare these items with the factors described in chapters 5 and 6.

Checklist A – Scoring Physical Design Factors

(1) Access (External): Determines if the external access is adequate for visual inspection and manipulative actions. Scoring will apply to external packaging as related to maintainability design concepts for ease of maintenance. This item is concerned with the design for external visual and manipulative actions which would precede internal maintenance actions. The following scores and scoring criteria will apply:

Scores
(a) Access adequate both for visual and manipulative tasks (electrical and mechanical)   4
(b) Access adequate for visual, but not manipulative, tasks   2
(c) Access adequate for manipulative, but not visual, tasks   2
(d) Access not adequate for visual or manipulative tasks   0

Scoring Criteria
An explanation of the factors pertaining to the above scores is consecutively shown. This procedure is followed throughout for other scores and scoring criteria.

(a) To be scored when the external access, while visual and manipulative actions are being performed on the exterior of the subassembly, does not present difficulties because of obstructions (cables, panels, supports, etc.).
(b) To be scored when the external access is adequate (no delay) for visual inspection, but not for manipulative actions. External screws, covers, panels, etc., can be located visually; however, external packaging or obstructions hinders manipulative actions (removal, tightening, replacement, etc.).
(c) To be scored when the external access is adequate (no delay) for manipulative actions, but not for visual inspections. This applies to the removal of external covers, panels, screws, cables, etc., which present no difficulties; however, their location does not easily permit visual inspection.
(d) To be scored when the external access is inadequate for both visual and manipulative tasks. External covers, panels, screws, cables, etc., cannot be easily removed nor visually inspected because of external packaging or location.

(2) Latches and Fasteners (External): Determines if the screws, clips, latches, or fasteners outside the assembly require special tools, or if significant time was consumed in the removal of such items. Scoring will relate external equipment packaging and hardware to maintainability design concepts. Time consumed with preliminary external disassembly will be proportional to the type of hardware and tools needed to release them and will be evaluated accordingly.

Scores
(a) External latches and/or fasteners are captive, need no special tools, and
     require only a fraction of a turn for release   4
(b) External latches and/or fasteners meet two of the above three criteria   2
(c) External latches and/or fasteners meet one or none of the above three
     criteria   0

Scoring Criteria
(a) To be scored when external screws, latches, and fasteners are:

     (1) Captive
     (2) Do not require special tools
     (3) Can be released with a fraction of a turn

Releasing a 'DZUS' fastener which requires a 90-degree turn using a standard
screw driver is an example of all three conditions.

(b) To be scored when external screws, latches, and fasteners meet two of the
     three conditions stated in (a) above. An action requiring an Allen wrench
     and several full turns for release shall be considered as meeting only one of
     the above requirements.
(c) To be scored when external screws, latches, and fasteners meet only one or
     none of the three conditions stated in (a) above.

(3) Latches and Fasteners (Internal): Determines if the internal screws, clips,
     fasteners or latches within the unit require special tools, or if significant time
     was consumed in the removal of such items. Scoring will relate internal
     equipment hardware to maintainability design concepts. The types of latches
     and fasteners in the equipment and standardisation of these throughout the
     equipment shall tend to affect the task by reducing or increasing required
     time to remove and replace them. Consider 'internal' latches and fasteners
     to be within the interior of the assembly.

Scores
(a) Internal latches and/or fasteners are captive, need no special tools, and
     require only a fraction of a turn for release   4
(b) Internal latches and/or fasteners meet two of the above three criteria   2
(c) Internal latches and/or fasteners meet one or none of the above three
     criteria   0

Scoring Criteria
(a) To be scored when internal screws, latches and fasteners are:

     (1) Captive
     (2) Do not require special tools
     (3) Can be released with a fraction of a turn

Releasing a 'DZUS' fastener which requires a 90-degree turn using a standard screwdriver would be an example of all three conditions.

(b) To be scored when internal screws, latches, and fasteners meet two of the three conditions stated in (a) above. A screw which is captive can be removed with a standard or Phillips screwdriver, but requires several full turns for release.

(c) To be scored when internal screws, latches, and fasteners meet one of three conditions stated in (a) above. An action requiring an Allen wrench and several full turns for release shall be considered as meeting only one of the above requirements.

Checklist B – Scoring Design Dictates – Facilities

The intent of this questionnaire is to determine the need for external facilities. Facilities, as used here, include material such as test equipment, connectors, etc., and technical assistance from other maintenance personnel, supervisor, etc.

(1) *External Test Equipment*: Determines if external test equipment is required to complete the maintenance action. The type of repair considered maintainably ideal would be one which did not require the use of external test equipment. It follows, then, that a maintenance task requiring test equipment would involve more task time for set-up and adjustment and should receive a lower maintenance evaluation score.

Scores
(a) Task accomplishment does not require the use of external test equipment   4
(b) One piece of test equipment is needed   2
(c) Several pieces (2 or 3) of test equipment are needed   1
(d) Four or more items are required   0

Scoring Criteria
(a) To be scored when the maintenance action does not require the use of external test equipment. Applicable when the cause of malfunction is easily detected by inspection or built-in test equipment.

(b) To be scored when one piece of test equipment was required to complete the maintenance action. Sufficient information was available through the use of one piece of external test equipment for adequate repair of the malfunction.

(c) To be scored when 2 or 3 pieces of external test equipment are required to complete the maintenance action. This type of malfunction would be complex enough to require testing in a number of areas with different test equipment.

(d) To be scored when four or more pieces of test equipment are required to complete the maintenance action. Involves an extensive testing requirement to locate the malfunction. This would indicate that a least maintainable condition exists.

(2) Connectors: Determines if supplementary test equipment requires special fittings, special tools, or adaptors to adequately perform tests on the electronic system or subsystem. During troubleshooting of electronic systems, the minimum need for test equipment adaptors or connectors indicates that a better maintainable condition exists.

Scores
(a) Connectors to test equipment require no special tools, fittings, or adaptors 4
(b) Connectors to test equipment require some special tools, fittings, or adaptors (less than two)   2
(c) Connectors to test equipment require special tools, fittings, and adaptors (more than two)   0

Scoring Criteria
(a) To be scored when special fittings or adaptors and special tools are not required for testing. This would apply to tests requiring regular test leads (probes or alligator clips) which can be plugged into or otherwise secured to the test equipment binding post.
(b) Applies when one special fitting, adaptor or tool is required for testing. An example would be if testing had to be accomplished using a 10 dB attenuator pad in series with the test set.
(c) To be scored when more than one special fitting, adaptor, or tool is required for testing. An example would be when testing requires the use of an adaptor and an r.f. attenuator.

(3) Jigs or Fixtures: Determines if supplementary materials such as block and tackle, braces, dollies, ladder, etc., are required to complete the maintenance action. The use of such items during maintenance would indicate the expenditure of a major maintenance time and pinpoint specific deficiencies in the design for maintainability.

Scores
(a) No supplementary materials are needed to perform task   4
(b) No more than one piece of supplementary material is needed to perform task   2
(c) Two or more pieces of supplementary material are needed   0

Scoring Criteria
(a) To be scored when no supplementary materials (block and tackle, braces,

dollies, ladder, etc.) are required to complete maintenance. Applies when the maintenance action consists of normal testings and the removal or replacement of parts or components can be accomplished by hand, using standard tools.

(b) To be scored when one supplementary material is required to complete maintenance. Applies when testing or when the removal and replacement of parts requires a step ladder for access or a dolly for transportation.

(c) To be scored when more than one supplementary material is required to complete maintenance. Concerns the maintenance action requiring a step ladder and dolly adequately to test and remove the replaced parts.

Checklist C – Scoring Design Dictates – Maintenance Skills

This check list evaluates the personnel requirements relating to physical, mental, and attitude characteristics, as imposed by the maintenance task.

Evaluation procedure for this check list can best be explained by way of several examples. Consider the first question which deals with arm, leg and back strength. Should a particular task require the removal of an equipment drawer weighing 100 pounds, this would impose a severe requirement on this characteristic. Hence, in this case the question would be given a low score (0 to 1). Assume another task which, due to small size and delicate construction, required extremely careful handling. Here question 1 would be given a high score (4), but the question dealing with eye-hand coordination and dexterity would be given a low score. Other questions in the check list relate to various personnel characteristics important to maintenance task accomplishment. In completing the check list, the task requirements for each of these characteristics should be viewed with respect to average technician capabilities.

Scores

	Score
1. Arm, leg, and back strength	____
2. Endurance and energy	____
3. Eye-hand coordination, manual dexterity, and neatness	____
4. Visual acuity	____
5. Logical analysis	____
6. Memory – things and ideas	____
7. Planfulness and resourcefulness	____
8. Alertness, cautiousness, and accuracy	____
9. Concentration, persistence and patience	____
10. Initiative and incisiveness	____

Scoring Criteria

Quantitative evaluations of these items range from 0 to 4 and are defined in the following manner:

4. The maintenance action requires a minimum effort on the part of the technician.
3. The maintenance action requires a *below average* effort on the part of the technician.
2. The maintenance action requires an *average* effort on the part of the technician.
1. The maintenance action requires an *above average* effort on his part.
0. The maintenance action requires a *maximum* effort on his part.

16.5 ANOTHER CHECKLIST METHOD

A paper entitled 'Development of Maintainability Prediction Methods Based Upon Checklists and Multiple Regression Techniques' was presented at the National Reliability Conference.

This was a result of the Reliability Research Group of The Admiralty Surface Weapons Establishment placing a study contract with The Plessey Company.

This has resulted in a computerised checklist method, similar to US MIL 472 Method 3, based on UK Naval equipment. Main differences are:

(i)   The scoring is against maintenance features of the equipment in general rather than for specific faults.

(ii) The method is subdivided into three categories of material:

> Electronic/Electro-mechanical
> Waveguide components
> Mechanical equipments

(iii) The three checklist groups are:

> Maintenance philosophy
> Location and access
> Diagnosis and test

(iv) A log Normal distribution of repair times is assumed.

Part IV
Essential Management Topics

17 Project Management

17.1 SETTING OBJECTIVES AND SPECIFICATIONS

Realistic reliability and maintainability objectives need to be set with due regard to the customer's design and operating requirements and cost constraints. Some discussion and joint study with the customer may be required to establish economic reliability values which sensibly meet his requirements and are achievable within the proposed technology at the costs allowed for. Over-specifying the requirement may delay the project when tests eventually show that objectives cannot be met and it is realised that budgets will be exceeded. When specifying an MTBF it is a common mistake to include a confidence level, in fact the MTBF requirement stands alone. The addition of a confidence level implies a demonstration and supposes that the MTBF would be established by a single demonstration at the stated confidence. On the contrary, a design objective is a target and must be stated without statistical limitations.

Consideration of the equipment type and the use to which it is put will influence the parameters chosen. Remember the comments in chapter 3 on failure rate, MTBF, Availability, MTTR, etc.

A major contribution to the problems of cost and quality comes from the lack of, or inadequacy of, the engineering design specification. It should specify the product requirements in full including reliability and MTTR parameters. These factors should include:

(i)	Functional Description	– Speeds, functions, human interfaces and operating periods.
(ii)	Environment	– Temperature, humidity, etc.
(iii)	Design Life	– Related to wearout and replacement policy.
(iv)	Physical Parameters	– Size and weight restrictions, power supply limits.
(v)	Standards	– BS, US MIL, Def Con, etc., standards for materials, components and tests.
(vi)	Finishes	– Appearance and materials.
(vii)	Ergonomics	– Human limitations and safety considerations.
(viii)	Reliability and Maintainability	– Module reliability and MTTR objectives. Equipment R and M related to module levels.

(ix) Manufacturing Quantity — Projected manufacturing levels — 1st off, Batch, Flow.

(x) Maintenance Philosophy — Type and frequency of preventive maintenance. Repair level, method of diagnosis, method of 2nd line repair.

17.2 PLANNING, FEASIBILITY AND ALLOCATION

The design and assurance activities described in this book will simply not occur in practice unless a reliability and maintainability programme is laid down and specific resources allocated. Responsibilities have to be placed on individuals for each of the activities and a reliability programme manager appointed with sufficient authority and the absence of conflicting priorities (that is programme dates) to control the R and M objectives. Milestones, with dates, will be required against which progress can be measured as, for example:

Completion of feasibility study (including R and M calculations)*.
Reliability objectives for modules and for bought out items allocated.
Test specification prepared and agreed.
Prototype tests completed.*
Modifications arising from tests completed.
Demonstrations of reliability and maintainability.
Design review dates (should include *).

The purpose of a feasibility study is to establish if the performance specification can be met within the constraints of cost, technology, time and so on. This involves a brief reliability prediction, based perhaps on a parts count approach, in order to decide if the design proposal has a reasonable chance of being engineered to meet the requirements. Allocation of objectives has been emphasised in chapter 4 and is important if the objectives are not to be met by a mixture of over and under-design.

17.3 PROGRAMME ACTIVITIES

The extent of the reliability and maintainability activities in a project will depend upon:

The severity of the requirement.
The complexity of the product.
Time and cost constraints.
Safety considerations.
The number of items to be produced.

These activities include:

Setting Objectives — Discussed above with allocation and feasibility.

Training — Design engineers should be trained to a level where they can work with the R and M specialist. Customer training of maintenance staff is another aspect which may arise.

Quality Assurance — This involves manufacturing controls to ensure correct materials, tolerances, etc., and all the activities of Quality Engineering, Test Planning, Test and Inspection, Reliability Engineering, etc.

Design Review — This is intended to provide an evaluation of the design at defined milestones. The design review board should comprise a variety of skills and be chaired by a person independent of the design team. The following checklist is a guide to the factors which might be considered:

(a) Electrical factors involving critical features, component standards, circuit trade-offs, etc.

(b) Software reliability involving configuration control, flowcharts, user documentation, etc.

(c) Mechanical features such as materials and finish, industrial design, ergonomics, equipment practice and so on.

(d) Quality and reliability covering environmental testing, predictions and demonstrations, FMECA, test equipment and procedures, trade-offs, etc.

(e) Maintenance philosophy including repair policy, MTTR prediction, maintenance resource forecasts, customer training and manuals.

(f) Purchased items involving lead times, multiple sourcing, supplier evaluation and make/buy decisions.

(g) Manufacturing and installation covering tolerances, burn in, packaging and transport, costs, etc.

(h) Other items include patents, value engineering, safety, documentation standards and product liability.

FMECA and Predictions — This focuses attention on the critical failure areas, highlights failures which are difficult to diagnose and provides a measure of the design reliability against the objectives.

Design Trade-Offs — These may be between R and M and may involve sacrificing one for the other as, for example, between the reliability of the wrapped joint and the easy replaceability of a connector. Major trade-offs will involve the design review whereas others will be made by the designer.

Prototype Tests — These cover marginal, functional, parametric, environmental and reliability tests. It is the first opportunity to observe reliability in practice and make some comparison against the predictions.

Parts Selection and Approval — Involves field tests or seeking field information from other users. The continued availability of each part is important and may influence the choice of supplier.

Spares Provisioning — This effects reliability and maintainability and has to be calculated during design.

Data Collection and Failure Analysis — Failure data, with the associated stress information, is essential to reliability growth programmes and also to future predictions. A formal failure reporting scheme should be set up at an early stage so that tests on the earliest prototype modules contribute towards the analysis.

Demonstrations — Since these involve statistical sampling test plans have to be calculated at an early stage so that the risks can be evaluated.

17.4 RESPONSIBILITIES

Reliability and maintainability are engineering design parameters and the responsibility for their achievement is primarily with the design team. Quality assurance techniques play a vital role in achieving the goals but cannot be used to 'test in' reliability to a design which has its own inherent level. Three distinct responsibilities therefore emerge which are complementary but do not replace each other.

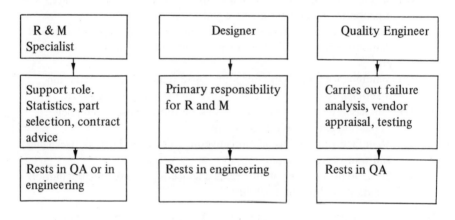

18 Contract Clauses and their Pitfalls

Since the late 1950s in the United States reliability and maintainability requirements have appeared in both military and civil engineering contracts. These contracts often carry penalties for failure to meet these objectives. For some years in the UK suppliers of military and commercial electronic and telecommunication equipment have also found that clauses specifying reliability and maintainability are being included in invitations to tender and in the subsequent contracts. Suppliers of highly reliable and maintainable equipment are often well able to satisfy such conditions with little or no additional design or manufacturing effort, but incur difficulty and expense since a formal demonstration of these parameters may not have been attempted before. Furthermore a failure reporting procedure may not exist and therefore historical data as to a product's reliability or repair time may be unobtainable. The inclusion of system effectiveness parameters in a contract involves both the suppliers of good and poor equipment in additional activities. System Effectiveness clauses in contracts range from a few words — specifying failure rate or MTBF of all or part of the system — to some ten or twenty pages containing details of design and test procedures, methods of collecting failure data, methods of demonstrating reliability and repair time, limitations on component sources, limits to size and cost of test equipment, and so on. Two types of pitfall arise from such contractual conditions.

(a) Those due to the omission of essential conditions or definitions.
(b) Those due to inadequately worded conditions which present ambiguities, concealed risks, eventualities unforeseen by both parties, etc.

The following headings are essential if reliability or maintainability is to be specified.

18.1.1 Definitions

If a mean time to repair is specified then the meaning of repair time must be explained in detail. It could be made up of many combinations of the elements discussed in chapter 3 or might even refer to down time or time when revenue is not being earned. Mean time to repair is often used when it is mean down time which is intended.

Failure itself must also be thoroughly defined at system and module levels. It may be necessary to define more than one type of failure (for example total system failure; degradation failure) or failures for different operating modes (for example in flight; on ground) in order to describe all the requirements. MTBFs might then be ascribed to the different failure types. MTBFs and failure rates often require clarification as to the meaning of 'failure' and 'time'. The latter may refer to operating time, revenue time, clock time, etc. Types of failure which do not count for the purpose of proving the reliability (for example maintenance induced, environment outside limits) have also to be defined. Figure 18.1 reminds us of the bathtub curve with early, random and wearout failures. Reliability parameters usually refer to random failures unless stated to the contrary, it being assumed that burn-in failures are removed by soaking and wearout failures eliminated by preventive replacement. It should be remembered that this is a statistical picture of the situation and that, in practice, it is rarely possible to ascribe a particular failure to any of these categories. It is therefore vital that, if reliability is being demonstrated by a test or in the field, these early and wearout failures are eliminated, as far as possible, by the measures already described. The specification should make clear which types of failure are being observed in a test.

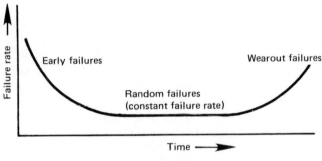

Figure 18.1

Parameters should not be used without due regard to their meaning and applicability. Failure rate, for example, has little meaning except when describing random failures. Remember that in systems involving redundancy constant failure rate may not apply except in the special cases outlined in chapter 15. MTBF or reliability should then be specified in preference.

Reliability and maintainability are often combined by specifying the useful parameter, availability. This can be defined in more than one way and should therefore be defined. The usual form is the Steady State Availability which is MTBF/(MTBF + MDT) where MDT is the Mean Down Time.

18.1.2 Environment

A common mistake is to fail to specify the environmental conditions under which the product is to work. The specification is often confined to temperature range and maximum humidity and this is not always adequate. Even these two parameters can create problems as with temperature cycling under high humidity conditions. Other stress parameters (chapter 8) include pressure, vibration and shock, chemical and bacteriological attack, power supply variations and interference, radiation, human factors and many others. The combination or the cycling of any of these parameters can have dramatic results.

Where equipments are used as standby units or held as spares the environmental conditions will be different to those experienced by operating units. It is often assumed that because a unit is not powered, or in store, it will not fail. In fact the environment may be more conducive to failure under these circumstances. Self-generated heat and mechanical self-cleaning wiping actions are often important ingredients for reliability. If equipment is to be transported whilst the supplier is liable for failure then the environmental conditions must be evaluated. On the other hand, overspecifying environmental conditions is a temptation for the customer which leads to over design and higher costs. Environmental testing is expensive, particularly if large equipments are involved and if vibration tests are called for. These costs should be quantified by obtaining quotations from a number of test houses before any commitment is made to demonstrate equipment under environmental conditions.

Maintainability can also be influenced by environment. Conditions relating to safety, comfort, health and ergonomic efficiency will influence repair times since the use of protective clothing, remote handling devices, safety precautions, etc., increases the active elements of repair time by slowing down the technician.

18.1.3 Maintenance Support

The provision of spares, test equipment, manpower, transport and the maintenance of both spares and test equipment is a responsibility which may be divided between supplier and customer or fall entirely on either. These responsibilities must be described in the contract and the supplier must be conscious of the risks involved in the customer not meeting his side of the bargain.

If the supplier is responsible for training the customer's maintenance staff then levels of skill and training have to be laid down.

Maintenance philosophy, usually under customer control, plays a part in determining reliability. Periodic inspection of a non-attended system during which failed redundant units are changed yields a different MTBF to the case of immediate repair of failed units irrespective of whether they result in system failure. The maintenance philosophy must therefore be defined.

A contract may specify an MTTR supported by a statement such as 'identification of faulty modules will be automatic and will be achieved by automatic test means. No additional test equipment will be required for diagnosis'. This type of requirement involves considerable additional design effort in order to permit all necessary diagnostic signals to be made accessible and for measurements to be made. Additional hardware will be required either in the form of BITE or an 'intelligent' portable terminal with diagnostic capability. If such a requirement is overlooked when costing and planning the design the subsequent engineering delay and cost is likely to be considerable.

18.1.4 Demonstration

The supplier might be called upon to give a statistical demonstration of either reliability or repair time. In the case of maintainability a number of corrective or preventive maintenance actions will be carried out and a given MTTR, or better, will have to be achieved for some proportion of the attempts. In this situation it is essential to define the tools and equipment to be used, the maintenance instructions, test environment and technician level. The method of task selection, the spares and the level of repair to be carried out also require stating. The probability of failing the test should be evaluated since some standard tests carry high supplier's risks. When reliability is being demonstrated then a given number of hours will be accumulated and a number of failures stated, above which the test is failed. Again, statistical risks apply and the supplier needs to calculate the probability of failing the test with good equipment and the customer that of passing inadequate goods. Essential parameters to define here are environmental conditions, allowable failures (for example maintenance induced), operating mode, preventive maintenance, burn-in, testing costs. It is often not possible to construct a reliability demonstration which combines sensible risks ($\leqslant 15$ per cent) for both parties with a reasonable length of test. Under these circumstances the acceptance of reliability may have to be on the basis of accumulated operating hours on previously installed similar systems. An alternative to statistical or historical demonstrations of repair time and reliability is a guarantee period wherein all or part of the failure costs, and sometimes redesign costs, are borne by the supplier. In these cases great care must be taken to calculate the likely costs. It must be remembered that if 100 equipments meet their stated MTBF under random failure conditions then after operating for a period equal to one MTBF 63 of them, on average, will have failed. From the point of view of producer's risk a warranty period is a form of reliability demonstration since, having calculated the expected number of failures during the warranty, there is a probability that more will occur. Many profit margins have been absorbed by the unbudgeted penalty maintenance arising from this fact.

18.1.5 Liability

The exact nature of the supplier's liability must be spelt out, including the maximum penalty which can be incurred. If some qualifying or guarantee period is involved it is necessary to define when this commences and when the supplier is free of liability. The borders between delivery, installation, commissioning and operation are often blurred and therefore the beginning of the guarantee period will be unclear.

It is wise to establish a mutually acceptable means of arbitration in case the interpretation of later events becomes the subject of a dispute. If part of the liability for failure or repair is to fall on some other contractor care must be taken in defining each party's area. The interface between equipment guaranteed by different suppliers may be physically easy to define but there exists the possibility of failures induced in one equipment due to failure or degraded performance in another. This point should be considered where more than one supplier is involved.

18.2 OTHER POSSIBLE AREAS

The following items are often covered in a detailed invitation to tender.

18.2.1 Reliability and Maintainability Programme

The detailed activities during design, manufacturing and installation are sometimes spelt out contractually. In a development contract this enables the customer to monitor the reliability and maintainability design activities and to measure progress against agreed milestones. Sometimes standard programme requirements are used as, for example;

> US MIL STD 470 – Maintainability Program Requirements.
> US MIL STD 785 – Requirements for Reliability Program.
> British Standard 4200 Part 5 – Reliability Programmes for Equipment.

Typical activities specified are:

Prediction – Data sources, mathematical models.
Testing – Methods and scheduling of design, environmental and other tests.
Design Review – Details of participation in design reviews.
Failure Mode and Effect Analysis – Details of method and timing.
Failure Reporting – Failure reporting documents and reporting procedures.

18.2.2 Reliability and Maintainability Analysis

The supplier may be required to offer a detailed reliability or maintainability prediction together with an explanation of the techniques and data used. Alternatively a prediction may be requested using defined data and methods of calculation. Insistence on optimistic data makes it more difficult to achieve the predicted values whereas pessimistic data leads to overdesign.

18.2.3 Storage

The equipment may be received by the customer and stored for some time before it is used under conditions different to normal operation. If there is a guarantee period then the storage conditions and durations will have to be defined. The same applies to storage and transport of spares and test equipment.

18.2.4 Design Standards

Specific design standards are sometimes described or referenced in contracts or their associated specifications. These can cover many areas including:

Printed board assemblies — design and manufacture
Wiring and soldering
Nuts, bolts and threads
Finishes
Component ratings
Packaging
etc., etc.

A problem exists that these standards are very detailed and most manufacturers have their own version. Although differences exist in the fine detail they are usually overlooked until some formal acceptance inspection takes place by which time retrospective action is difficult, time consuming and costly.

18.3 PITFALLS

The foregoing lists those aspects of reliability and maintainability likely to be mentioned in an invitation to tender or in a contract. There are pitfalls associated with the omission or inadequate definition of these factors and some of the more serious are outlined below.

18.3.1 Definitions

The most likely area of dispute is the definition of what constitutes a failure and whether or not a particular incident ranks as one or not. There are levels of failure (system, unit), types of failure (catastrophic, degradation), causes of failure (random, systematic, overstress) and there are effects of failure (dormant, hazardous). For various combinations of these, different MTBF and MTTR objectives with different penalties may be set. It is seldom sufficient, therefore, to define failure as not performing to specification since there are so many combinations covered by that statement. Careful definition of the failure types covered by the contract is therefore important.

18.3.2 Repair Time

It was shown in chapter 3 that repair times could be divided into elements. Initially they can be grouped into active and passive elements and, broadly speaking, the active elements are dictated by system design and the passive by maintenance and operating arrangements. For this reason the supplier should never guarantee any part of the repair time which is influenced by the user.

18.3.3 Statistical

A statistical maintainability test is described by a number of repair actions and an objective MTTR which must not be exceeded on more than a given number of attempts. A reliability test involves a number of hours and a similar pass criterion of a given number of failures. In both cases producer and consumer risks apply as explained in earlier chapters and unless these risks are calculated they can prove to be unacceptable. Where published test plans are quoted it is never a bad thing to recalculate the risks involved. It is not difficult to find a test which requires the supplier to achieve an MTBF 50 times the value which is to be proved in order to stand a reasonable chance of passing the test.

18.3.4 Quoted Specifications

Sometimes a reliability or maintainability programme or test plan is specified by calling up a published standard. Definitions are also sometimes dealt with this way. The danger with blanket definitions lies in the possibility that not all the quoted terms are suitable and that the standards will not be studied in every detail.

18.3.5 Environment

Environmental conditions effect both reliability and repair times. Temperature and humidity are the most usual to be specified and the problem of cycling has already been pointed out. If other factors are likely to be present in field use then they must either be specifically excluded from the range of environment for which the product is guaranteed or included and therefore allowed for in the design and in the price. It is not desirable to specify every parameter possible since this leads to overdesign.

18.3.6 Liability

When stating the supplier's liability it is important to establish its limit in terms of both cost and time. The supplier must ensure that he knows when he is finally free of liability.

18.3.7 In Summary

The biggest pitfall of all is to assume that either party wins any advantage from ambiguity or looseness in the conditions of a contract. In practice the manhours of investigation and negotiation which ensue from a dispute far outweigh any advantage that might have been secured, to say nothing of the loss of goodwill and reputation. If every effort is made to cover all the areas discussed as clearly and simply as possible then both parties will gain.

18.4 PENALTIES

There are various ways in which a penalty may be imposed on the basis of maintenance costs or the cost of system outage. Some alternatives are briefly outlined.

18.4.1 Apportionment of Costs During Guarantee

Figure 18.2a illustrates the method where the supplier pays the total cost of corrective maintenance during the guarantee period. He may also be liable for the cost of redesign made necessary by systematic failures. In some cases the guarantee period recommences for those parts of the equipment effected by modifications. A disadvantage of this arrangement is that it gives the customer no great incentive to minimise maintenance costs until the guarantee has expired. If the maintenance is carried out by the customer and paid for by the supplier then the latter's control over the preventive maintenance effectiveness is minimal. The customer should never be permitted to benefit from poor maintenance for which reason this method is not very desirable.

An improvement of this is obtained by figure 18.2b whereby the supplier pays a proportion of the costs during the guarantee and both parties therefore have an incentive to minimise costs. In figure 18.2c the supplier's proportion of the costs decreases over the liability period. In figure 18.2d the customer's share of the maintenance costs remains constant and the supplier pays the excess. The arrangements in (b) and (c) both provide mutual incentives. Arrangement (d) however provides a mixed incentive. The customer has, initially, a very high incentive to reduce maintenance costs but once the ceiling has been reached this disappears. On the other hand (d) recognises the fact that for a specified MTBF the customer should anticipate a given amount of repair. Above this amount the supplier pays for the difference between the achieved and contracted values.

18.4.2 Payment According to Down Time

The above arrangements involve penalties related to the cost of repair. Some contracts, however, demand a payment of some fixed percentage of the contract price during the down time. Providing that the actual sum paid is less than the

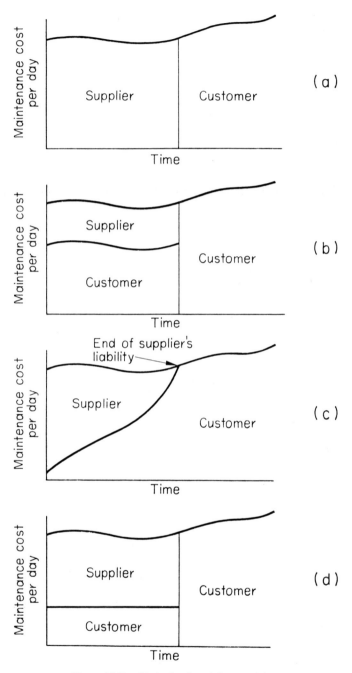

Figure 18.2   Methods of applying penalties

cost of the repair this method is similar to figure 18.2b although in practice it is not likely to be so generous. In any case an arrangement of this type must be subject to an upper limit.

18.4.3 In Summary

Except in case (a) it would not be practicable for the supplier to carry out the maintenance. Usually the customer carries out the repairs and the supplier pays according to some agreed rate. In this case he must require some control over the recording of repair effort and a right to inspect the customer's maintenance records and facilities from time to time. It should be remembered that achievement of reliability and repair time objectives does not imply zero maintenance costs. If a desired **MTBF** of 20 000 h is achieved for each of ten equipments then in one year (8760 h) about four failures can be expected. On this basis (d) is fairer than (a). When part of a system is subcontracted to another supplier then the prime contractor must ensure that he passes on an appropriate allocation of the system effectiveness commitments in order to protect himself.

19 Product Liability

Product liability is the liability of a supplier, designer or manufacturer to the customer for injury or loss resulting from a defect in that product. There are two main reasons why it has recently become the focus of attention. The first is the recent publication of a draft directive by the European Economic Community and the second is the wave of actions under United States Law which have resulted in spectacular awards for claims involving death or injury. In 1977 the average sum awarded resulting from court proceedings was $256 000. Changes in the United Kingdom are inevitable but it is first necessary to review the current position.

19.1 THE EXISTING SITUATION

19.1.1 Contract Law

This is largely governed by the Sale of Goods Act, 1893, which requires that goods are of merchantable quality and are reasonably fit for the purpose intended. Privity of contract exists between the buyer and seller which means that only the buyer has any remedy for injury or loss and then only against the seller. This is modified slightly by the Supply of Goods (Implied Terms) Act, 1973, which makes exclusion clauses void for consumer contracts. This means that a condition excluding the seller from liability would be void in law. Note that a contract does not have to be in writing and that a sale, in this context, implies the existence of a contract.

19.1.2 Common Law

The relevant area is that relating to the Tort of Negligence for which a claim for damages can be made. Everyone has a duty of care to his neighbour, in law, and failure to exercise reasonable precautions with regard to one's skill, knowledge and the circumstances involved constitutes a breach of that care. A claim for damages for common law negligence is, therefore, open to anyone and not restricted as in Privity of Contract. On the other hand the onus is with the plaintiff to prove negligence which requires him to prove:

That the product was defective.

That the defect was the cause of the injury.
That this was foreseeable and that the plaintiff failed in his duty of care.

19.1.3 Statute Law

The main Acts relevant to this area are:

Supply of Goods (Implied Terms) Act, 1973.
    Exclusion clauses void for consumer contracts.
    Exclusion clauses in business contracts only valid if considered reasonable.
Unfair Contract Terms Act, 1977.
    Courts can rule on what is considered reasonable.
    Scope includes services and hire purchase.
Consumer Safety Act, 1978.
    Gives official powers to require warning labels, to make prohibition orders
    against unsafe products, to make regulations and standards for fitness of
    purpose, etc.
Health and Safety at Work Act, 1974. Section 6.
    Involves the criminal law. Places a duty to construct and instal items, processes
    and materials without health or safety risks. It applies to places of work.
    Responsibility involves everyone including management.

19.1.4 In Summary

The present situation involves a form of strict liability but:

Privity of Contract excludes third parties.
The onus is to prove negligence.
Exclusion clauses, involving death and personal injury are voidable.

19.2 STRICT LIABILITY

19.2.1 Concept

The concept of strict liability hinges on the idea that liability exists for no other
reason than the mere existence of a defect. No breach of contract, or act of
negligence, is required in order to incur responsibility and a manufacturer will be
liable for compensation if his product causes injury.

The various recommendations which are summarised later involve slightly
different interpretations of strict liability ranging from the extreme case of
everyone in the chain of distribution and design being strictly liable to the
manufacturer being liable unless he can prove that the defect did not exist when
the product left him. Whereas the Law Commission urges a shift in the onus of
proof the EEC directive would make the manufacturer liable whether or not he
could have known of a potential hazard.

19.2.2 Defects

A defect, for the purposes of product liability, covers:

Manufacturing   — Presence of impurities or foreign bodies.
                       — Fault or failure due to manufacturing or installation.
Design   — Product not fit for the purpose stated.
                       — Inherent safety hazard in the design.
Documentation — Lack of necessary warnings.
                       — Inadequate or incorrect operating and maintenance instructions resulting in a hazard.

19.3 TRENDS AND RECOMMENDATIONS

19.3.1 Recommendations of the Law Commissions

Both English and Scottish Law Commissions presented a report in 1977 which in both cases recommended a form of strict liability. In general their recommendations include:

Strict liability but only for personal injury.
Liability is only avoided if the product is misused.
A defect exists if current standards are not complied with.
Responsibility extends to everyone in the chain of design and production.
The Scottish Commission alone thought that liability should cease when a
    product becomes a component in another product.

19.3.2 The Royal Commission (The Pearson Report)

This was published in 1978 and made similar recommendations to the Law Commissions. It concerned itself, not only with the law, but with methods of compensation including insurance and state benefits. The major points are:

Strict liability extending to suppliers of components.
Distributors are strictly liable.
Importers are strictly liable.
No power to contract out of liability.
A defence would be that injury was totally due to misuse and not caused by the
    defect.
Proceedings would have to be brought within 3 years of the awareness of
    damage.

19.3.3 EEC Draft Directive and The Strasbourg Convention

Both recommend strict liability, even without fault, as already discussed.

Principle differences between the two are:

EEC Draft Directive	*Strasbourg Convention*
Ignores contributory negligence	Recognises contributory negligence
Embraces property	Personal injuries only
Limit to compensation	No monetary limit to compensation.

19.4 HEALTH AND SAFETY AT WORK ACT, 1974

19.4.1 Scope

Section 6 of this Act is, in fact, strict liability applied to articles produced for use at work. It is very wide and embraces designers, manufacturers, suppliers, hirers and employers of industrial plant and equipment. We are now dealing with criminal law and failure to observe the duties laid down in the Act are punishable by fine or imprisonment. Claims for compensation are still dealt with in civil law.

19.4.2 Duties

The main items are:

To design and construct products without risk to health or safety.
To test according to the specification laid down.
To provide adequate information to the user for safe operation.
To make positive tests to evaluate risks and hazards.
To use safe methods of installation.
To use safe (proven) substances and materials.

19.4.3 Concessions

The main concessions are:

It is a defence that a product has been used without regard to the relevant information.
It is a defence that the design was carried out on the basis of a written undertaking by the purchaser to take specified steps to ensure the safe use of the item.
One's duty is restricted to matters within one's control.
One is not required to repeat tests upon which it is reasonable to rely.

19.4.4 Responsibilities

Basically everyone concerned in the design and provision of an article is responsible for it. Directors and Managers are held responsible for the designs and manufactured articles of their companies and are expected to take steps to assure safety in their products. Employees are also responsible. The 'buck' cannot be passed in either direction.

19.5 INSURANCE

19.5.1 The Effect of Product Liability Trends

An increase in the number of claims.
Higher premiums.
The creation of separate Product Liability Policies.
Involvement of insurance companies in defining quality and reliability standards and procedures.

19.5.2 Some Critical Areas

All Risks — This means all risks specified in the policy. Check that your requirements are met by the policy.
Comprehensive — Essentially means the same as the above.
Disclosure — The policy holder is bound to disclose any information relevant to the risk. Failure to do so, whether asked for or not, can invalidate a claim. The test of what should be disclosed is described as 'anything the prudent insurer should know'.
Exclusions — The Unfair Contract Terms Act, 1977 does not apply to insurance so read and negotiate accordingly. For example defects related to design could be excluded and this would considerably weaken a policy from the product liability standpoint.

19.5.3 Areas of Cover

Premiums are usually expressed as a percentage of turnover and cover is divided into three areas:

Product Liability — Cover against claims for personal injury or loss.
Product Guarantee — Cover against the expenses of warranty/repair.
Product Recall — Cover against the expenses of recall.

19.6 PRODUCT RECALL

19.6.1 Types of Recall

A design defect causing a potential hazard to life, health or safety may become evident when a number of products are already in use. It may then become necessary to recall, for replacement or modification, a batch of items some of which may be spread throughout the chain of distribution and others in use. The recall may vary in the degree of urgency depending on whether the hazard is to life, health or merely reputation. A hazard which could reasonably be thought to endanger life or to create a serious health hazard should be treated by an emergency recall procedure. Where less critical risks involving minor health and safety hazards are discovered a slightly less urgent approach may suffice. A third category, operated at the vendor's discretion, applies to defects causing little or no personal hazard and where only reputation is at risk.

If it becomes necessary to implement a recall the extent will be determined by the nature of the defect. It might involve, in the worst case, every user or maybe only a specific batch of items. In some cases the modification may be possible in the field and in others physical return of the item will be required. In any case a full evaluation of the hazard must be made and a report prepared.

19.6.2 Implementing the Recall

One person, usually the Quality Manager, must be responsible for the handling of the recall and he must be directly answerable to the Managing Director or Chief Executive. The first task is to prepare, if appropriate, a Hazard Notice in order to warn those likely to be exposed to the risk. Circulation may involve individual customers when traceable, field service staff, distributors, or even the media. It will contain sufficient information to describe the nature of the hazard and the precautions to be taken. Instructions for returning the defective item can be included, preferably with a prepaid return card. Small items can be returned with the card whereas large ones, or products to be modified in the field, will be retained whilst arrangements are made.

Where products are despatched to known customers a comparison of returns with output records will enable a 100 per cent check to be made on the coverage. Where products have been despatched in batches to wholesalers or retail outlets the task is not so easy and the quantity of returns can only be compared with a known output, perhaps by area. Individual users cannot be traced with 100 per cent certainty. Where customers have completed and returned record cards after purchase the effectiveness of the recall is improved.

After the recall exercise has been completed a major investigation into the causes of the defect must be made and the results progressed through the company's Quality and Reliability Programme. Causes could include:

Insufficient test hours.
Insufficient test coverage.
Insufficient information sought on materials.
Insufficient industrial engineering of the product prior to manufacture.
Insufficient production testing.
Insufficient field/user trials.
Insufficient user training.

20 A Case Study

THE DATAMET PROJECT

This chapter is a case study which has been used by the author, on Reliability and Maintainability Management and contract courses for nearly 10 years. It is not intended to represent any actual company, product or individuals.

The page entitled 'Syndicate Study' suggests a number of areas for thought and discussion. When discussing the contract clauses two syndicates can assume the two roles of producer and customer respectively. After individual study and discussion the two syndicates can renegotiate the contract under the guidance of the course tutor. This approach has proved both stimulating and effective. It is worth reflecting, when criticising the contract reliability clauses, that although the case study is fictional the clauses were drawn from actual examples.

20.1 INTRODUCTION

The Communications Division of ELECTROSYSTEMS Ltd has an annual turnover of £15 000 000. Current year's sales are forecast as follows:

	Line Communications	h.f. Radio	Special Systems
Home sales	£9 600 000	£2 000 000	£300 000
Export	£ 900 000	£ 900 000	£1 200 000

Line communications systems include 12 circuit, 4 MHz and 12 MHz multiplex systems. A highly reliable range of h.f. radio products includes ship-to-shore, radio beacons, SOS equipment, etc. Special systems constitute 10 per cent of sales and consist of equipment for transmitting information from oil wells and pipe lines over line systems.

The structure of the Division, which employs 1000 personnel, is shown in appendix 20.1 and that of the Engineering Department in appendix 20.2.

20.2 THE DATAMET CONCEPT

In June 1978 the Marketing Department was investigating the market potential for a meteorological telemetry system (DATAMET) whereby a number of observations at some remote location could be scanned, in sequence, and the information relayed by v.h.f. radio to a terminal station. Each observation is converted to an analogue signal in the range 0–10 V and up to 14 instruments can be scanned 4 times in one minute. Each signal in turn is used to frequency modulate a v.h.f. carrier. Several remote stations could operate on different carrier frequencies and, at the terminal, the remote stations are separated out and their signals interpreted and recorded.

An overseas administration showed an interest in purchasing 10 of these systems each to relay meteorological readings from 10 unattended locations. A total contract price of £1 500 000 for the 100 remote and the 10 terminal stations was mentioned. Marketing felt that some £6 000 000 of sales could be obtained for these systems over 5 years.

20.3 FORMATION OF THE PROJECT GROUP

The original feasibility group consisted of Peter Kenton (Special Systems section head), Len Ward (Radio Lab section head) who had some v.h.f. experience and Arthur Parry (a sales engineer).

A suggested design involved the continuous transmission of each reading on a different frequency. This was found to be a costly solution and, since continuous monitoring was not essential, a scanning system was proposed. appendix 20.3 illustrates the system whereby each instrument reading is converted to an electrical analogue in the 0–10 V range. The 14 channels are scanned by a microprocessor controller which sends each signal in code form, to the modulator unit. Each remote station operates at a different frequency in the region of 30 MHz. After each cycle of 14 signals a synchronising signal, outside the normal voltage range, is sent. The terminal station consists of a receiver and demodulator for separating out the remote stations. The signal from each station is then divided into 14 channels and fed to a desk top calculator with printer.

A meteorological equipment supplier was found who was prepared to offer instruments converting each reading to a 0–10 V signal. Each set of 14 instruments would cost £1400 for the quantities involved.

Due to the interest shown by the potential overseas customer it was decided to set up a project group with Kenton as Project Manager. The group consisted of Ward and another radio engineer, 2 special systems engineers, 3 equipment engineers and 4 technicians. The project organisation, with Kenton reporting to Ainsworth, is shown in appendix 20.4. In September 1978 Kenton prepared the project plan shown in appendix 20.5.

20.4 RELIABILITY REQUIREMENTS

In week 5 the customer expressed a firm intention to proceed and the following requirements became known:

Remote stations
MTBF of 5 years
Preventive maintenance at 6 month intervals
Equipment situated in windproof hut with inside temperature range 0–50 °C
Cost of corrective maintenance for the first year to be borne by supplier

Terminal
MTBF of 2000 h
Maximum repair time of 1 h

The first of the 10 systems was to be installed by week 60 and the remainder at 1 month intervals. The penalty maintenance clause was to take effect, for each station, at the completion of installation.

The customer produced a draft contract in week 8 and Parry was asked to evaluate the reliability clauses which are shown in appendix 20.6.

20.5 FIRST DESIGN REVIEW

The first design review was chaired by Ainsworth and took place in week 10. It consisted of Kenton, Parry, Ward, Jones, Ainsworth and the Marketing Manager. Kenton provided the following information:

From appendix 20.5 the project group would expend 250 manweeks.
Engineering assistance would be 70 manweeks for Drawing, Model Shop, Test
    equipment building, technical writing.
All engineering time was costed at £400 per manweek.
The parts for the laboratory model would cost £10 000.
The production model which would involve 1 terminal and 2 remote stations
    would cost £60 000.
Likely production cost for the systems would be £100 000 for a terminal with
    10 remotes. The above costs did not include the instruments.

On the basis of these costs the project was considered satisfactory if a minimum of four such contracts was to be received.

An initial crude reliability prediction had been carried out by Kenton for the remote equipment and this is reproduced in appendix 20.7. It assumed random failures, generous component tolerancing, commercial components and fixed ground conditions. A multiplication factor of 1.5 was applied to the data to allow for the rather more stringent conditions and a Mean Time Between Failures of about 4 years was obtained. Since no redundancy had been assumed this represented a worst case estimate and Kenton maintained that the objective of 5 years would eventually be met. Ward, however, felt that the factor of 1.5 was quite inadequate since the available data referred to much more controlled conditions. A factor of 3 would place Kenton's estimate nearly an order below the objective and he therefore held that more attention should be given to reliability at this stage. He was overruled by Kenton who was extremely optimistic about the whole project.

The outline design was agreed and it was recorded that attention should be given to:

(a) The LSI devices.
(b) Obtaining an MTBF commitment from the instrument supplier.
(c) Thorough laboratory testing.

20.6 DESIGN AND DEVELOPMENT

The contract, for £1 500 000, was signed in week 12 with 2 modifications to the reliability section. Parry insisted that the maximum of 1 h for repair should be replaced by a mean time to repair of 30 min since it is impossible to guarantee a maximum repair time. For failures to the actual instruments the labour costs were excluded from the maintenance penalty. Purchasing obtained a 2 000 000 h MTBF commitment from the instrument supplier.

Design continued and by week 20 circuits were being tested and assembled into a laboratory model. Kenton carried out a second reliability prediction in week 21 taking account of some circuit redundancy and of the 6 monthly visits. Ward still maintained that a multiplication factor of 3 was needed and Kenton agreed to a compromise by using 2.5. This yielded an MTBF of 7 years for a remote station. Ward pointed out that even if an MTBF of 8 years was observed in practice then, during the first year, some 12 penalty visits could be anticipated. The cost of a repair involving an unscheduled visit to a remote station could well be in the order of £1200.

At the commencement of laboratory testing Ward produced a failure reporting format and suggested to Parry that the customer should be approached concerning its use in field reporting. Since a maintenance penalty had been accepted he felt that there should be some control over the customer's failure reporting. In the meantime the format was used during laboratory testing and Ward was disturbed to note that the majority of failures arose from combinations of drift conditions rather than from catastrophic component failures. Such failures in the field would be likely to be in addition to those anticipated from the predicted MTBF.

In week 30 the supplier of the instruments became bankrupt and it was found that only 6 sets of instruments could be supplied. With some difficulty, an alternative supplier was found who could provide the necessary instruments. Modifications to the system were required since the new instruments operated over a 0–20 V range. The cost was £1600 per set of 14.

20.7 SYNDICATE STUDY

First Session

1. Comment on the Project Plan prepared by Kenton.
   (a) What activities were omitted, wrongly timed or inadequately performed?
   (b) How would you have organised this project?
2. Comment on the organisation of the project group.
   (a) Do you agree with the reporting levels?
   (b) Were responsibilities correctly assigned?
3. Is this project likely to be profitable? If not in what areas is money likely to be lost?

Second Session

1. Discuss the contract clauses and construct alternatives either as
   (i) Producer
   (ii) Customer
2. Set up a role playing negotiation.

20.8 HINTS

1. Consider the project, and projected figures, as percentage of turnover.
2. Compare the technologies in the proposed design with the established product range and look for differences.
3. Look for the major sources of failure (rate).
4. Consider the instrument reliability requirement and the proposed sourcing.
5. Think about appraisal of the design feasibility.
6. This book has frequently emphasised Allocation.
7. Why is this not a development contract?
8. How were responsibilities apportioned?
9. Were appropriate parameters chosen? (Availability).
10. What were the design objectives?
11. Think about test plans and times.
12. Schedule Design Reviews.
13. Define failure modes and types with associated requirements.

APPENDIX 20.1 STRUCTURE OF THE DIVISION

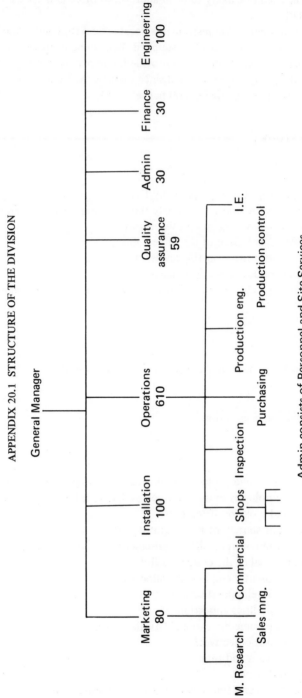

Admin consists of Personnel and Site Services
Numbers indicate approximate numbers of personnel

APPENDIX 20.2 ENGINEERING DEPARTMENT

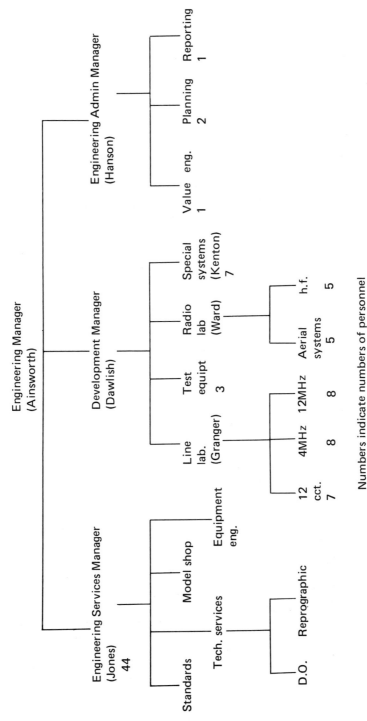

Numbers indicate numbers of personnel

APPENDIX 20.3 THE DATAMET SYSTEM

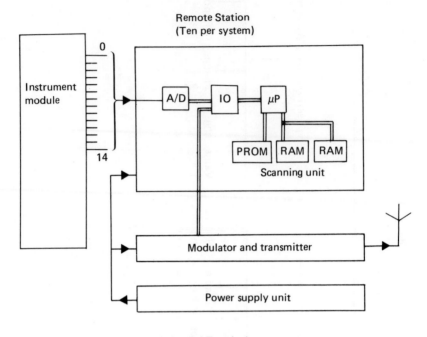

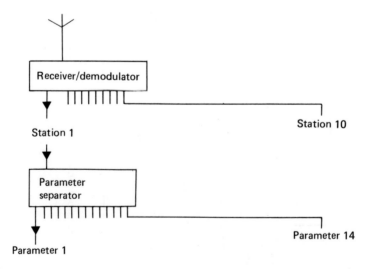

APPENDIX 20.4 PROJECT ORGANISATION

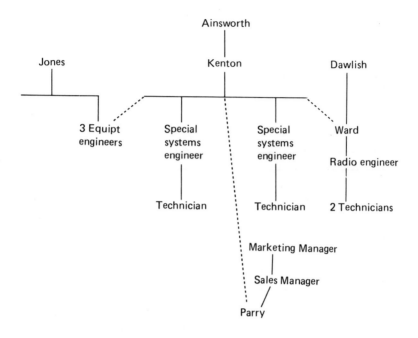

APPENDIX 20.5 PROJECT PLAN

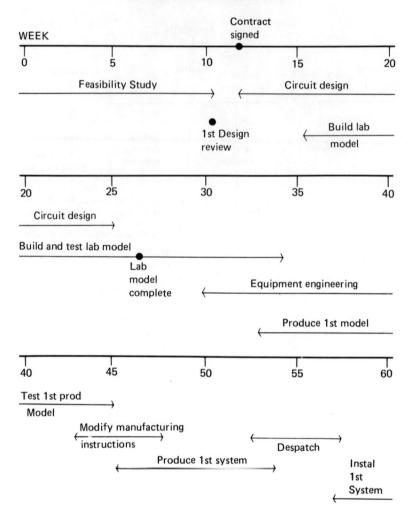

APPENDIX 20.6 CONTRACT RELIABILITY CLAUSES

(a) Five years mean time between failures is required for each remote station, 2000 h mean time between failures for the terminal. The supplier will satisfy the customer, by means of a reliability prediction, that the design is capable of meeting these objectives.

(b) The equipment must be capable of operating in a temperature range of 0–50 °C with a maximum relative humidity of 80 per cent.

(c) Failure shall consist of the loss of any parameter or the incorrect measurement of any parameter.

(d) For one year's operation of the equipment the contractor will refund the cost of all replacements to the terminal equipment and to the remote equipment. Where a corrective maintenance visit, other than a routine visit, is required the contractor shall refund all labour and travelling costs including overtime and incentives at a rate to be agreed.

(e) In the event of a system failure then the maximum time to restore the terminal to effective operation shall be 1 h. The contractor is required to show that the design is compatible with this objective.

(f) In the event of systematic failures the contractor shall perform all necessary redesign work and make the necessary modifications to all systems.

(g) The contractor is to use components having the most reasonable chance of being available throughout the life of the equipment and is required to state shelf life and number of spares to be carried in the case of any components that might cease to be available.

(h) The use of interchangeable printed cards may be employed and a positive means of identifying which card is faulty must be provided so that, when a fault occurs, it can be rectified with the minimum effort and skill. The insertion of cards in the wrong position shall be impossible or shall not cause damage to the cards or system.

(i) Maintenance instructions will be provided by the contractor and shall contain all necessary information for the checking and maintenance of the system. These shall be comprehensive and give full operational and functional information. The practice of merely providing a point to point component description of the circuits will not, in itself, be adequate.

APPENDIX 20.7 RELIABILITY PREDICTION

	Number	λ	$k\lambda$	$Nk\lambda$	
Instruments	14	500	—	7000	
Connections	14	2	3	42	
					7042
Cyclic Switch					
Microprocessor	1	400	600	600	
Memory chips	3	400	600	1800	
Other chips	2	500	750	1500	
Capacitors	15	150	225	3375	
Transistors	15	60	90	1350	
Solder joints	250	1	1.5	375	
Pins	50	2	3	150	
					9150
Modulator and Transmitter					
Varactors	1	1000	1500	1500	
Transistors	10	40	60	600	
Resistors	30	8	12	360	
Trimmers	3	140	210	630	
Capacitors	12	80	120	1400	
Crystal	1	1000	1500	1500	
Transformer	1	300	450	450	
Solder joints	150	1	1.5	225	
Pins	20	2	3	60	
					6725
Power					
Transformer	1	1000	1500	1500	
Transistors	10	60	90	900	
Zeners	3	140	210	630	
Power diodes	4	100	150	600	
Capacitors (electrolytic)	6	200	300	1800	
Solder joints	40	1	1.5	60	
Pins	10	2	3	30	
					5520

$$28437 \times 10^{-9}/\text{h}$$

Therefore MTBF = 35 000 h = 4 years

21 Software and Reliability

21.1 THE EFFECT OF PROGRAMMABLE DEVICES ON RELIABILITY

There has been a spectacular growth during the 1970s in the use of programmable devices. These are generally described as microprocessors and they have made a significant impact on methods of electronic circuit design. The main effect has been to reduce the number of different circuit types by the use of computer architecture coupled with software programming which provides the individual circuit features previously achieved by differences in hardware. The word software refers to any programme needed to enable a computer type device to function. This development of programming at the circuit level, now common with most industrial and consumer products, brings with it the associated quality and reliability problems. When applied to microprocessors at the circuit level the programming which is semi-permanent and usually contained in ROM (Read Only Memory) is known as Firmware. The necessary increase in function density of devices in order to provide the large quantities of memory in small packages has matched this trend. Computing and its associated software, is seen in three broad categories:

Mainframe computing — Isolated processing of large quantities of data and no interaction with real time events. Known as 'data crunching'.

Minicomputing — Interactive processing where real time events are monitored and control of peripheral devices is provided. Many examples, including Automatic Test Equipment, Process Control, Medical Equipment, etc. As size of minis increases some 'data crunching' tasks are catered for.

Microprocessing — Single chip computer functions used in electronic circuit design to provide functions previously achieved with hard wired logic. Known as embedded computing — where the computing hardware is not separable from the product itself.

From the quality and reliability point of view, both advantages and disadvantages exist with the programmable design solutions.

Reliability Advantages	*Reliability Disadvantages*
Less hardware (fewer devices) per circuit.	Difficult to 'inspect' software for errors.
Fewer device types.	Difficult to impose standard approaches to software design.
Consistent architecture (configuration).	Difficult to control software changes.
Common approach to hardware design.	Testing of LSI devices difficult due to high package density and
Easier to support several models (versions) in the field.	therefore reduced interface with test equipment.

The question arises as to how a software failure is defined. Unlike hardware there is no physical change associated with a unit that is 'functional' at one moment and 'failed' at the next. Software failures are in fact errors which, due to the complexity of a computer program, do not become evident until some moment when the combination of conditions brings the error to light. The effect is then the same as any other failure. Unlike the hardware bathtub there is no wearout characteristic but only a continuing burn-in. Each time that a change to the software is made the error rate is likely to rise as shown in figure 21.1.

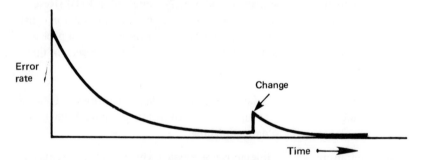

Figure 21.1   Software error curve

As a result of software errors there has been, for some time, an interest in developing methods of controlling the activities of programmers and of reducing software complexity by attempts at standardisation.

21.2 DOCUMENTATION CONTROLS

Documentation is an essential part of engineering and software is no exception. The first step towards standardisation is to establish a system of software documentation and then to provide controls so that it is used. This provides a formal discipline but does not, of itself, ensure error free programming. It simply provides a workable framework with which to carry out checks. Standard formats are developed and all concerned must be involved in their use. These will include:

(i) *Software Functional Specifications*

Describe the functions to be performed, the messages to be sent, message format, etc.

(ii) *Flowcharts*

Logically describe the program in block diagram algorithmic form as shown in the example of figure 21.2. Flow diagrams are the first step in the production of a program and should contain notes recording the programmer's thoughts at the time. The example shows an element of a control program where some parameter Y is compared with a required value X. Whilst $Y < X$, Y is incremented step by step until $Y = X$.

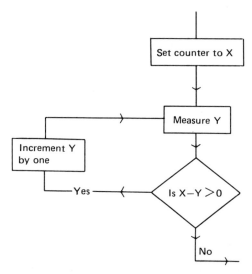

Figure 21.2   Flowchart

(iii) *D-Charts*

This is an improved method of 'flowcharting' a structured program. It consists of downward flowing lines, with symbols replacing the traditional boxes. The structured definitions implicit in the flow charts are part of the language, hence making coding easier.

(iv) *Listings*

The listing, with comment, is generated in assembly or high level language as the programmer produces program steps from the flow chart.

(v) *Acceptance Tests*

Consist of detailed descriptions of how programs are to be checked. Usually a functional test procedure is produced which covers all the permutations of functions for which the product is designed. In this way it is assumed that all possible routes in the flow chart are verified if the functional test is passed. In practice, programs are sufficiently complex that some combinations of events can be overlooked when writing the test plan and hence functional tests do not guarantee error free software.

(vi) *Change Documentation (figure 21.3)*

As with hardware the need to ensure that changes are documented and correctly applied to all media and program documents is vital. All programs

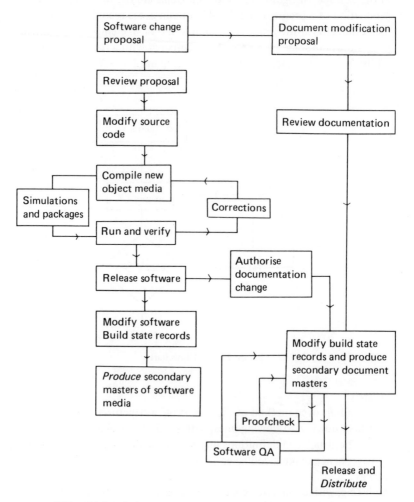

Figure 21.3   Software change and documentation procedure

and their associated documents should therefore carry issue numbers as do equipments. A formal document and software change procedure is required so that all change proposals are reviewed for their effect on the total system. It is essential that documents relate to the software (on tapes, discs, memory, etc.) at all times. This is even more important than with hardware, since the visibility to the program is only through the documents, or by inference from the system performance. Unlike hardware there is no easy visual check that the two are identical.

(vii) *Hardware Configuration Control*

Any change to the hardware configuration can cause the software not to be compatible. The two have therefore to be closely co-ordinated.

(viii) *Error Reporting*

All errors and discrepancies should be recorded. A precise description of the occurrence together with the issue status and the circumstances is required. Each incident must be investigated and the change procedure invoked as required.

21.3 PROGRAMMING STANDARDS AND CONTROLS

21.3.1 Structured Programming

The aim of structured programming is to reduce program complexity by using a library of defined structures wherever possible. The human brain is not well adapted to retaining random information and sets of standard rules and concepts substantially reduce the likelihood of error. A standard approach to creating files, polling output devices, handling interrupt routines and so on, constrains the programmer to use the proven methods. The use of specific sub-routines is a further step in this direction. Once a particular sequence of program steps has been developed in order to execute a specific calculation, then it should be used as a library sub-routine by the rest of the team. Re-inventing the wheel is both a waste of time and an unecessary source of failure if an error free program has already been developed.

21.3.2 Control of Coding

A set of written instructions, Perhaps in Manual form, should define for the programmer:

The language and Instruction Set.
The Documents to be produced (e.g. flow diagram, listings, etc.).
Style and layout of the listing.
Standard program format.
Symbols to be used in flowcharts.
QA rules for checking.

Change procedure.
Production of media (tapes, discs).
History files and available sub-routines.
Design aids as available.

Comments should be written along with and at the same time as the coding, and explain what the program is doing.

21.4 TESTING OF SOFTWARE

21.4.1 Proprietary Software

In just the same way as it has to be established, for purchased hardware, that adequate testing has been carried out, the same checks should be made for software. An alternative source of confidence is the existence of supporting field usage data, relating to the appropriate hardware environment. In the absence of either of these then further testing may have to be carried out. Subcontracted software must be subject to the same QA techniques as that produced in-house and the subcontractor required to demonstrate adequate controls and checks. Subcontracted hardware is sometimes handed over before the total system, of which it is part, has been fully developed. In this case only a rigorous check of the software against the interface specification is possible. This must be particularly thorough since no system tests can be carried out.

21.4.2 In-House Software

A good test plan, as well as good data, is needed. In addition to exercising the modules with the test data the results of the tests must be verified for accuracy Any changes, and the reasons for them, should be recorded in detail. The main areas of Software QA are:

(i) *Documentation Review*
   Programmer's documentation includes
   Complete problem definition and specification.
   Algorithm and data structure descriptions.
   Description of the logic and structure of the program.
   A clean, complete set of program listings.
   Definitive statement of testing and verification procedures.
   Operating instructions: start-up procedures, error messages, etc.

   User's documentation includes
   Condensed problem statement.
   What the program does and why.

Program logic and listings may be omitted.

Test results are included only if informative to the user.

Operating instructions in their entirety.

The package should meet each requirement listed below.

A set of charts showing the functional flow of the system should be included.

The specification gives representative names and mnemonics for the system and individual modules ('meaningful names').

The functions that the system and the modules are expected to perform have been clearly defined.

There is a description of the performance and capabilities required of each module and the system.

Timing requirements are given for both run time and real time.

Limitations affecting testing and/or operation of the software are defined.

If software must be operated within some vehicle or support equipment constraints, the facilities and their degrees of involvement are described.

Computer memory requirements (where the software will reside in memory) and peripheral requirements are stated.

A description of the operating and executive system requirements is included.

The support and utility software requirements are given for test software.

Software functions requiring human reactions and operator responses are defined.

A description of the loading configuration and type of control required is included.

Describe messages:

— Description of output

— Types of messages

— Format

— How they are transmitted

Describe input/output quantity, format, type of data and method. Give the test configuration, showing hardware and software required.

List test data required for verification tests, showing the form and format.

Identify the method of verification to be employed and list the capabilities to be verified.

(ii) Code Checking and Proof of Correctness.

This is essentially a documentation inspection exercise. It involves checking the coding (the written statements in high level or assembly language) against the specification and flow chart in order to find discrepancies. The flow chart algorithm can also be checked against the requirements specification for logical errors. This check is often called a walkthrough and the benefits are:

'Bugs' in the code are detected earlier and can therefore be more easily and cheaply corrected.

Inefficiencies in the code may be detected.

The walkthrough provides an ideal opportunity for training new personnel in programming standards.

Since more than one person is involved in the walkthrough the strengths and weaknesses of individual programmers may be complemented.

During the walkthrough at least one other programmer becomes familiar with the program thereby providing security and backup.

The group effort involved tends to generate a team attitude which encourages cross fertilisation of ideas.

(iii) Software Proving by Emulation.

An 'intelligent' communications analyser or other simulator having programmable stimulus and response facilities is used to emulate parts of the system not yet developed. In this way the software can be made to interact with the emulator which appears as if it were the surrounding hardware and software.

Software testing can thus proceed before the total system is complete.

(iv) System Functional Testing.

The ultimate empirical test is to assemble the system and to make it carry out every possible function as described in 21.2.(v). This is described by a complex test procedure and should cover the full range of environmental conditions specified.

(v) System Load Testing.

The situation may exist where a computer controls a number of smaller microprocessors, data channels or even hard wired equipments. The full quantity of these peripheral devices may not be available during test particularly if the system is designed for expansion. In these cases, it is necessary to simulate the full number of inputs by means of a simulator. A further micro or minicomputer may well be used for this purpose. Test software will then have to be written which emulates the total number of devices and sends and receives data from the processor under test. This is similar to the emulation described in (iii) but involves repetitive messages at the necessary data rate to simulate a fully loaded system. In most cases the processor both sends and receives data and commands in which case the configuration of figure 21.4 is used in order to check:

(a) The processor can receive data correctly at full load.

(b) The processor can send data and commands correctly at full load.

21.5 DATA COMMUNICATIONS

A further source of errors, and hence system failures, is the random bit errors introduced by the data communications network including lines, modems,

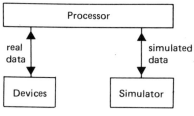

Figure 21.4   Load test

multiplexors and switching. A data bit error rate of 10^{-6} means that 1 binary bit in 10^6 bits will be incorrect as a result of line noise or interference. This does not necessarily mean that 1 message in 10^6 will be corrupted since it is normal to send additional binary bits containing parity or other error checking codes. For example a 112 bit message might contain 96 bits of data and 16 bits of coded information derived from the other 96. A comparision of these 16 bits (when received) with another 16 bits similarly calculated from the received data permits an error check to be performed. If the two error codes are not identical the data is retransmitted. In this way only a small fraction of the random bit errors will pass undetected by the check.

Errors are cumulative and a system involving several data links, multiplexors and so on can be treated as a series reliability problem with random error (failure) rates ascribed to each section and the error rates added after suitable calculations to take account of the error codes on each link. This type of error checking is analogous to the use of redundancy for improving hardware reliability.

Figure 21.5 shows a communications system involving a controlling minicomputer connected, via a multiplexor, to a number of lines. A peripheral device is connected to each line and contains a magnetic read/write unit which produces magnetically encoded cards.

Let the error bit rates for each item in this chain be defined as:

x_1 for those errors introduced by the   minicomputer

x_2 for those errors introduced by the   multiplexor

x_3 for those errors introduced by the   line

x_4 for those errors introduced by the   microprocessor in the peripheral

x_5 for those errors introduced by the   read/write unit and its connection to the microprocessor unit

Between the processor in the peripheral and the minicomputer each message is 112 bits in length but contains 16 bits of error check code as already described.

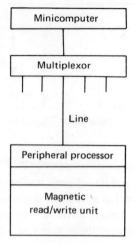

Figure 21.5

The number of wrongly coded cards per bit error in the appropriate part of the system is reduced by a factor of:

$$\frac{96}{112} \cdot \frac{1}{2^{16}} = \frac{96}{112} \cdot \frac{1}{65536} = \frac{1}{76458}$$

Between the magnetic unit and the processor the 96 bits of information are passed with no additional error checking.

The error rate of wrongly coded cards is therefore:

$$96x_5 + \frac{(x_1 + x_2 + x_3 + x_4)}{76\,458}$$

Appendix 1 Glossary

A1. TERMS RELATED TO FAILURE

A1.1 Failure

Termination of the ability of an item to perform its specified function. OR, Non-conformance to some defined performance criteria. Failures may be classified by:

Meaningless without performance spec.

(i) Cause — Misuse: Caused by operation outside specified stress.
Primary: Not caused by an earlier failure.
Secondary: Caused by an earlier failure.
Wearout: Caused by accelerating failure rate mechanism.
Design: Caused by an intrinsic weakness.
Software: Caused by a program error despite no hardware failure.

chapter 21

(ii) Type — Sudden: Not anticipated and no prior degradation.
Degradation: Parametric drift or gradual reduction in performance.
Intermittent: Alternating between the failed and operating condition.
Dormant: A component or unit failure which does not cause system failure but which either hastens it or, in combination with another dormant fault, would cause system failure.
Random: Failure is equally probable in each successive equal time interval.
Catastrophic: Sudden and complete.

A1.2 Failure Mode

The outward appearance of a failure.

A1.3 Failure Mechanism

The physical or chemical process which causes the failure.   chapter 7

A1.4 Failure Rate

The number of failures of an item per unit time.   Per hour, cycle,
                                                          operation, etc.

   This can be applied to:

  (i) Observed failure rate: as computed from a sample.   Point estimate
 (ii) Assessed failure rate: as inferred from sample   Involves a
     information.   confidence level
(iii) Extrapolated failure rate: projected to other stress
     levels.

A1.5 Mean Time Between Failures and Mean Time to Fail
The total cumulative functioning time of a population divided by the number of
failures. As with failure rate the same applies to Observed, Assessed and
Extrapolated MTBF. MTBF is used for items which involve repair. MTTF is used
for items with no repair.

A2. RELIABILITY TERMS

A2.1 Reliability

The probability that an item will perform a required
function, under stated conditions, for a stated period of
time.
   Since observed reliability is empirical it is defined
as the ratio of items which perform their function for the
stated period to the total number in the sample.

A2.2 Redundancy

The provision of more than one means of achieving a   Replication
function.
   Active: All items remain operating prior to failure.
   Standby: Replicated items do not operate until needed.

A2.3 Reliability Growth

Increase in reliability as a result of continued design
modifications resulting from field data feedback.

A2.4 Intrinsic (Inherent) Reliability

The basic reliability level dictated by the design and components and taking no account of failures added by manufacture, installation, wearout or operation.

A3. MAINTAINABILITY TERMS

A3.1 Maintainability

The probability that a failed item will be restored to operational effectiveness within a given period of time when the repair action is performed in accordance with prescribed procedures.

A3.2 Mean Time to Repair

The mean time to carry out a defined maintenance action.   Usually refers to corrective maintenance

A3.3 Repair Rate

The reciprocal of MTTR.

When used in reliability calculations it is the reciprocal of Down Time

A3.4 Repair Time

The time during which an item is undergoing diagnosis, repair, checkout and alignment.

Must be carefully defined

A3.5 Down Time

The time during which an item is not able to perform to specification.

Must be carefully defined

A3.6 Corrective Maintenance

The actions associated with repair time.

A3.7 Preventive Maintenance

The actions, other than corrective maintenance,
carried out for the purpose of keeping an item in a
specified condition.

A3.8 Least Replaceable Assembly

That assembly at which diagnosis ceases and   Typically a
replacement is carried out.   printed board
                                                     assembly

A3.9 Second Line Maintenance

Maintenance of LRAs which have been removed from
the field for repair or for preventive maintenance.

A4. TERMS ASSOCIATED WITH SOFTWARE

A4.1 Software

All documentation and inputs (for example tapes, discs)
associated with programmable devices.

A4.2 Programmable Device

Any piece of equipment containing one or more components
which provides a computer architecture with memory facilities.

A4.3 High Level Language

A means of writing program instructions using symbols
which each represent several program steps.

A4.4 Assembler

A program for converting program instructions, written in
mnemonics, into binary machine code suitable to operate a
programmable device.

A4.5 Compiler

A program which, in addition to being an assembler,
generates more than one instruction for each statement
thereby permitting the use of a high level language.

A4.6 Diagnostic Software

A program containing self-test algorithms enabling
failures to be identified.

Particularly
applicable to ATE

A4.7 Simulation

The process of representing a unit or system by some
means in order to provide some or all identical inputs, at
some interface, for test purposes.

A4.8 Emulation

A type of simulation whereby the simulator responds
to all possible inputs as would the real item and
generates all the corresponding outputs.

Identical to the real
item from the point
of view of a unit
under test

A4.9 Load Test

A system test involving simulated inputs in order to
prove that the system will function at full load.

A4.10 Functional Test

An empirical test routine designed to exercise an item
such that all aspects of the software are brought into use.

A4.11 Software Error

An error in the programming causing a malfunction in use.

A4.12 Bit Error Rate

The random incidence of incorrect binary digits.

Expressed
10^{-x}/bit

A4.13 Automatic Test Equipment (ATE)

Equipment for stimulus and measurement controlled by a
programmed sequence of steps (usually in software).

A5 MISCELLANEOUS TERMS

A5.1 Availability (Steady State)

The proportion of time that an item is capable of   Given as:
operating to specification within a large time interval.   MTBF/MTBF + MDT

A5.2 Burn-In

The operation of items for a specified period of time in order to remove early failures and bring the reliability characteristic into the random failure part of the bathtub curve.

A5.3 Confidence Interval

A range of a given variable within which a random value will lie at a stated confidence (probability).

A5.4 Consumer's Risk

The probability of an unacceptable batch being accepted due to a favourable sample.

A5.5 Derating

The use of components having a higher strength rating in order to reduce failure rate.

A5.6 Ergonomics

The study of man/machine interfaces in order to minimise human errors due to mental or physical fatigue.

A5.7 Mean

Usually used to indicate the Arithmetic Mean which is the sum of a number of values divided by the number thereof.

A5.8 Median

The median is that value such that 50 per cent of the values in question are greater and 50 per cent less than it.

A5.9 Producer's Risk

The probability of an acceptable batch being rejected
due to an unfavourable sample.

A5.10 Quality

Conformance to specification.

A5.11 Random

Such that each item has the same probability of being
selected as any other.

A5.12 System Effectiveness

A general term covering the subject involving
Availability, Reliability and Maintainability.

A5.13 Terotechnology

As integrated approach to the overall optimisation of
life cycle costs and resources.

Appendix 2 Percentage Points of the χ^2 Distribution

n \ α	0·9995	0·999	0·995	0·990	0·975	0·95	0·90	0·80	0·70	0·60
1	0.0^6393	0.0^5157	0.0^4393	0.0^3157	0.0^3982	0.0^2393	0·0158	0·0642	0·148	0·275
2	0.0^2100	0.0^2200	0·0100	0·0201	0·0506	0·103	0·211	0·446	0·713	1·02
3	0·0153	0·0243	0·0717	0·115	0·216	0·352	0·584	1·00	1·42	1·87
4	0·0639	0·0908	0·207	0·297	0·484	0·711	1·06	1·65	2·19	2·75
5	0·158	0·210	0·412	0·554	0·831	1·15	1·61	2·34	3·00	3·66
6	0·299	0·381	0·676	0·872	1·24	1·64	2·20	3·07	3·83	4·57
7	0·485	0·598	0·989	1·24	1·69	2·17	2·83	3·82	4·67	5·49
8	0·710	0·857	1·34	1·65	2·18	2·73	3·49	4·59	5·53	6·42
9	0·972	1·15	1·73	2·09	2·70	3·33	4·17	5·38	6·39	7·36
10	1·26	1·48	2·16	2·56	3·25	3·94	4·87	6·18	7·27	8·30
11	1·59	1·83	2·60	3·05	3·82	4·57	5·58	6·99	8·15	9·24
12	1·93	2·21	3·07	3·57	4·40	5·23	6·30	7·81	9·03	10·2
13	2·31	2·62	3·57	4·11	5·01	5·89	7·04	8·63	9·93	11·1
14	2·70	3·04	4·07	4·66	5·63	6·57	7·79	9·47	10·8	12·1
15	3·11	3·48	4·60	5·23	6·26	7·26	8·55	10·3	11·7	13·0
16	3·54	3·94	5·14	5·81	6·91	7·96	9·31	11·2	12·6	14·0
17	3·98	4·42	5·70	6·41	7·56	8·67	10·1	12·0	13·5	14·9
18	4·44	4·90	6·26	7·01	8·23	9·39	10·9	12·9	14·4	15·9
19	4·91	5·41	6·84	7·63	8·91	10·0	11·7	13·7	15·4	16·9
20	5·40	5·92	7·43	8·26	9·59	10·9	12·4	14·6	16·3	17·8
21	5·90	6·45	8·03	8·90	10·3	11·6	13·2	15·4	17·2	18·8
22	6·40	6·98	8·64	9·54	11·0	12·3	14·0	16·3	18·1	19·7
23	6·92	7·53	9·26	10·2	11·7	13·1	14·8	17·2	19·0	20·7
24	7·45	8·08	9·98	10·9	12·4	13·8	15·7	18·1	19·9	21·7
25	7·99	8·65	10·5	11·5	13·1	14·6	16·5	18·9	20·9	22·6
26	8·54	9·22	11·2	12·2	13·8	15·4	17·3	19·8	21·8	23·6
27	9·09	9·80	11·8	12·9	14·6	16·2	18·1	20·7	22·7	24·5
28	9·66	10·4	12·5	13·6	15·3	16·9	18·9	21·6	23·6	25·5
29	10·2	11·0	13·1	14·3	16·0	17·7	19·8	22·5	24·6	26·5
30	10·8	11·6	13·8	15·0	16·8	18·5	20·6	23·4	25·5	27·4
31	11·4	12·2	14·5	15·7	17·5	19·3	21·4	24·3	26·4	28·4
32	12·0	12·8	15·1	16·4	18·3	20·1	22·3	25·1	27·4	29·4
33	12·6	13·4	15·8	17·1	19·0	20·9	23·1	26·0	28·3	30·3
34	13·2	14·1	16·5	17·8	19·8	21·7	24·0	26·9	29·2	31·3
35	13·8	14·7	17·2	18·5	20·6	22·5	24·8	27·8	30·2	32·3
36	14·4	15·3	17·9	19·2	21·3	23·3	25·6	28·7	31·1	33·3
37	15·0	16·0	18·6	20·0	22·1	24·1	26·5	29·6	32·1	34·2
38	15·6	16·6	19·3	20·7	22·9	24·9	27·3	30·5	33·0	35·2
39	16·3	17·3	20·0	21·4	23·7	25·7	28·2	31·4	33·9	36·2
40	16·9	17·9	20·7	22·2	24·4	26·5	29·1	32·3	34·9	37·1
41	17·5	18·6	21·4	22·9	25·2	27·3	29·9	33·3	35·8	38·1
42	18·2	19·2	22·1	23·7	26·0	28·1	30·8	34·2	36·8	39·1
43	18·8	19·9	22·9	24·4	26·8	29·0	31·6	35·1	37·7	40·0
44	19·5	20·6	23·6	25·1	27·6	29·8	32·5	36·0	38·6	41·0
45	20·1	21·3	24·3	25·9	28·4	30·6	33·4	36·9	39·6	42·0
46	20·8	21·9	25·0	26·7	29·1	31·4	34·2	37·8	40·5	43·0
47	21·5	22·6	25·8	27·4	30·0	32·3	35·1	38·7	41·5	43·9
48	22·1	23·3	26·5	28·2	30·8	33·1	35·9	39·6	42·4	44·9
49	22·8	24·0	27·2	28·9	31·6	33·9	36·8	40·5	43·4	45·9
50	23·5	24·7	28·0	29·7	32·4	34·8	37·7	41·4	44·3	46·9

0·50	0·40	0·30	0·20	0·10		0·05	0·025	0·01	0·005	0·001		0·0005	α / n
0·455	0·708	1·07	1·64	2·71		3·84	5·02	6·63	7·88	10·8		12·1	1
1·39	1·83	2·41	3·22	4·61		5·99	7·38	9·21	10·6	13·8		15·2	2
2·37	2·95	3·67	4·64	6·25		7·81	9·35	11·3	12·8	16·3		17·7	3
3·36	4·04	4·88	5·99	7·78		9·49	11·1	13·3	14·9	18·5		20·0	4
4·35	5·13	6·06	7·29	9·24		11·1	12·8	15·1	16·7	20·5		22·1	5
5·35	6·21	7·23	8·56	10·6		12·6	14·4	16·8	18·5	22·5		24·1	6
6·35	7·28	8·38	9·80	12·0		14·1	16·0	18·5	20·3	24·3		26·0	7
7·34	8·35	9·52	11·0	13·4		15·5	17·5	20·1	22·0	26·1		27·9	8
8·34	9·41	10·7	12·2	14·7		16·9	19·0	21·7	23·6	27·9		29·7	9
9·34	10·5	11·8	13·4	16·0		18·3	20·5	23·2	25·2	29·6		31·4	10
10·3	11·5	12·9	14·6	17·3		19·7	21·9	24·7	26·8	31·3		33·1	11
11·3	12·6	14·0	15·8	18·5		21·0	23·3	26·2	28·3	32·9		34·8	12
12·3	13·6	15·1	17·0	19·8		22·4	24·7	27·7	29·8	34·5		36·5	13
13·3	14·7	16·2	18·2	21·1		23·7	26·1	29·1	31·3	36·1		38·1	14
14·3	15·7	17·3	19·3	22·3		25·0	27·5	30·6	32·8	37·7		39·7	15
15·3	16·8	18·4	20·5	23·5		26·3	28·8	32·0	34·3	39·3		41·3	16
16·3	17·8	19·5	21·6	24·8		27·6	30·2	33·4	35·7	40·8		42·9	17
17·3	18·9	20·6	22·8	26·0		28·9	31·5	34·8	37·2	42·3		44·4	18
18·3	19·9	21·7	23·9	27·2		30·1	32·9	36·2	38·6	43·8		46·0	16
19·3	21·0	22·8	25·0	28·4		31·4	34·2	37·6	40·0	45·3		47·5	20
20·3	22·0	23·9	26·2	29·6		32·7	35·5	38·9	41·4	46·8		49·0	21
21·3	23·0	24·9	27·3	30·8		33·9	36·8	40·3	42·8	48·3		50·5	22
22·3	24·1	26·0	28·4	32·0		35·2	38·1	41·6	44·2	49·7		52·0	23
23·3	25·1	27·1	29·6	33·2		36·4	39·4	43·0	45·6	51·2		53·5	24
24·3	26·1	28·2	30·7	34·4		37·7	40·6	44·3	46·9	52·6		54·9	25
25·3	27·2	29·2	31·8	35·6		38·9	41·9	45·6	48·3	54·1		56·4	26
26·3	28·2	30·3	32·9	36·7		40·1	43·2	47·0	49·6	55·5		57·9	27
27·3	29·2	31·4	34·0	37·9		41·3	44·5	48·3	51·0	56·9		59·3	28
28·3	30·3	32·5	35·1	39·1		42·6	45·7	49·6	52·3	58·3		60·7	29
29·3	31·3	33·5	36·3	40·3		43·8	47·0	50·9	53·7	59·7		62·2	30
30·3	32·3	34·6	37·4	41·4		45·0	48·2	52·2	55·0	61·1		63·6	31
31·3	33·4	35·7	38·5	42·6		46·2	49·5	53·5	56·3	62·5		65·0	32
32·2	34·4	36·7	39·6	43·7		47·4	50·7	54·8	57·6	63·9		66·4	33
33·3	35·4	37·8	40·7	44·9		48·6	52·0	56·1	59·0	65·2		67·8	34
34·3	36·5	38·9	41·8	46·1		49·8	53·2	57·3	60·3	66·6		69·2	35
35·3	37·5	39·9	42·9	47·2		51·0	54·4	58·6	61·6	68·0		70·6	36
36·3	38·5	41·0	44·0	48·4		52·2	55·7	59·9	62·9	69·3		72·0	37
37·3	39·6	42·0	45·1	49·5		53·4	56·9	61·2	64·2	70·7		73·4	38
38·3	40·6	43·1	46·2	50·7		54·6	58·1	62·4	65·5	72·1		74·7	39
39·3	41·6	44·2	47·3	51·8		55·8	59·3	63·7	66·8	73·4		76·1	40
40·3	42·7	45·2	48·4	52·9		56·9	60·6	65·0	68·1	74·7		77·5	41
41·3	43·7	46·3	49·5	54·1		58·1	61·8	66·2	69·3	76·1		78·8	42
42·3	44·7	47·3	50·5	55·2		59·3	63·0	67·5	70·6	77·4		80·2	43
43·3	45·7	48·4	51·6	56·4		60·5	64·2	68·7	71·9	78·7		81·5	44
44·3	46·8	49·5	52·7	57·5		61·7	65·4	70·0	73·2	80·1		82·9	45
45·3	47·8	50·5	53·8	58·6		62·8	66·6	71·2	74·4	81·4		84·2	46
46·3	48·8	51·6	54·9	59·8		64·0	67·8	72·4	75·7	82·7		85·6	47
47·3	49·8	52·6	56·0	60·9		65·2	69·0	73·7	77·0	84·0		86·9	48
48·3	50·9	53·7	57·1	62·0		66·3	70·2	74·9	78·2	85·4		88·2	49
49·3	51·9	54·7	58·2	63·2		67·5	71·4	76·2	79·5	86·7		89·6	50

n＼α	0·9995	0·999	0·995	0·990	0·975	0·95	0·90	0·80	0·70	0·60	0·50
51	24·1	25·4	28·7	30·5	33·2	35·6	38·6	42·4	45·3	47·8	50·3
52	24·8	26·1	29·5	31·2	34·0	36·4	39·4	43·3	46·2	48·8	51·3
53	25·5	26·8	30·2	32·0	34·8	37·3	40·3	44·2	47·2	49·8	52·3
54	26·2	27·5	31·0	32·8	35·6	38·1	41·2	45·1	48·1	50·8	53·3
55	26·9	28·2	31·7	33·6	36·4	39·0	42·1	46·0	49·1	51·7	54·3
56	27·6	28·9	32·5	34·3	37·2	39·8	42·9	47·0	50·0	52·7	55·3
57	28·2	29·6	33·2	35·1	38·0	40·6	43·8	47·9	51·0	53·7	56·3
58	28·9	30·3	34·0	35·9	38·8	41·5	44·7	48·8	51·9	54·7	57·3
59	29·6	31·0	34·8	36·7	39·7	42·3	45·6	49·7	52·9	55·6	58·3
60	30·3	31·7	35·5	37·5	40·5	43·2	46·5	50·6	53·8	56·6	59·3
61	31·0	32·5	36·3	38·3	41·3	44·0	47·3	51·6	54·8	57·6	60·3
62	31·7	33·2	37·1	39·1	42·1	44·9	48·2	52·5	55·7	58·6	61·3
63	32·5	33·9	37·8	39·9	43·0	45·7	49·1	53·5	56·7	59·6	62·3
64	33·2	34·6	38·6	40·6	43·8	46·6	50·0	54·3	57·6	60·5	63·3
65	33·9	35·4	39·4	41·4	44·6	47·4	50·9	55·3	58·6	61·5	64·3
66	34·6	36·1	40·2	42·2	45·4	48·3	51·8	56·2	59·5	62·5	65·3
67	35·3	36·8	40·9	43·0	46·3	49·2	52·7	57·1	60·5	63·5	66·3
68	36·0	37·6	41·7	43·8	47·1	50·0	53·5	58·0	61·4	64·4	67·3
69	36·7	38·3	42·5	44·6	47·9	50·9	54·4	59·0	62·4	65·4	68·3
70	37·5	39·0	43·3	54·4	48·8	51·7	55·3	59·9	63·3	66·4	69·3
71	38·2	39·8	44·1	46·2	49·6	52·6	56·2	60·8	64·3	67·4	70·3
72	38·9	40·5	44·8	47·1	50·4	53·5	57·1	61·8	65·3	68·4	71·3
73	39·6	41·3	45·6	47·9	51·3	54·3	58·0	62·7	66·2	69·3	72·3
74	40·4	42·0	40·4	48·7	52·1	55·2	58·9	63·6	67·2	70·3	73·3
75	41·1	42·8	47·2	49·5	52·9	56·1	59·8	64·5	68·1	71·3	74·3
76	41·8	43·5	48·0	50·3	53·8	56·9	60·7	65·5	69·1	72·3	75·3
77	42·6	44·3	48·8	51·1	54·6	57·8	61·6	66·4	70·0	73·2	76·3
78	43·3	45·0	49·6	51·9	55·5	58·7	62·5	67·3	71·0	74·2	77·3
79	44·1	45·8	50·4	52·7	56·3	59·5	63·4	68·3	72·0	75·2	78·3
80	44·8	46·5	51·2	53·5	57·2	60·4	64·3	69·2	72·9	76·2	79·3
81	45·5	47·3	52·0	54·4	58·0	61·3	65·2	70·1	73·9	77·2	80·3
82	46·3	48·0	52·8	55·2	58·8	62·1	66·1	71·1	74·8	78·1	81·3
83	47·0	48·8	53·6	56·0	59·7	63·0	67·0	72·0	75·8	79·1	82·3
84	47·8	49·6	54·4	56·8	60·5	63·9	67·9	72·9	76·8	80·1	83·3
85	48·5	50·3	55·2	57·6	61·4	64·7	68·8	73·9	77·7	81·1	84·3
86	49·3	51·1	56·0	58·5	62·2	65·6	69·7	74·8	78·7	82·1	85·3
87	50·0	51·9	56·8	59·3	63·1	66·5	70·6	75·7	79·6	83·0	86·3
88	50·8	52·6	57·6	60·1	63·9	67·4	71·5	76·7	80·6	84·0	87·3
89	51·5	53·4	58·4	60·9	64·8	68·2	72·4	77·6	81·6	85·0	88·3
90	52·3	54·2	59·2	61·8	65·6	69·1	73·3	78·6	82·5	86·0	89·3
91	53·0	54·9	60·0	62·6	66·5	70·0	74·2	79·5	83·5	87·0	90·3
92	53·8	55·7	60·8	63·4	67·4	70·9	75·1	80·4	84·4	88·0	91·3
93	54·5	56·5	61·6	64·2	68·2	71·8	76·0	81·4	85·5	88·9	92·3
94	55·3	57·2	62·4	65·1	69·1	72·6	76·9	82·3	86·4	89·9	93·3
95	56·1	58·0	63·2	65·9	69·9	73·5	77·8	83·2	87·3	90·9	94·3
96	56·8	58·8	64·1	66·7	70·8	74·4	78·7	84·2	88·3	91·9	95·3
97	57·6	59·6	64·9	67·6	71·6	75·3	79·6	85·1	89·2	92·9	96·3
98	58·4	60·4	65·7	68·4	72·5	76·2	80·5	86·1	90·2	93·8	97·3
99	59·1	61·1	66·5	69·2	73·4	77·0	81·4	87·0	91·2	94·8	98·3
100	59·9	61·9	67·3	70·1	74·2	77·9	82·4	87·9	92·1	95·8	99·3

0·40	0·30	0·20	0·10	0·05	0·025	0·01	0·005	0·001	0·0005	α
										n
52·9	55·8	59·2	64·3	68·7	72·6	77·4	80·7	88·0	90·9	51
53·9	56·8	60·3	65·4	69·8	73·8	78·6	82·0	89·3	92·2	52
55·0	57·9	61·4	66·5	71·0	75·0	79·8	83·3	90·6	93·5	53
56·0	58·9	62·5	67·7	72·2	76·2	81·1	84·5	91·9	94·8	54
57·0	60·0	63·6	68·8	73·3	77·4	82·3	85·7	93·2	96·2	55
58·0	61·0	64·7	69·9	74·5	78·6	83·5	87·0	94·5	97·5	56
59·1	62·1	65·7	71·0	75·6	79·8	84·7	88·2	95·8	98·8	57
60·1	63·1	66·8	72·2	76·8	80·9	86·0	89·5	97·0	100·1	58
61·1	64·2	67·9	73·3	77·9	82·1	87·2	90·7	98·3	101·4	59
62·1	65·2	69·0	74·4	79·1	83·3	88·4	92·0	99·6	102·7	60
63·2	66·3	70·0	75·5	80·2	84·5	89·6	93·2	100·9	104·0	61
64·2	67·3	71·1	76·6	81·4	85·7	90·8	94·4	102·2	105·3	62
65·2	68·4	72·2	77·7	82·5	86·8	92·0	95·6	103·4	106·6	63
66·2	69·4	73·3	78·9	83·7	88·0	93·2	96·9	104·7	107·9	64
67·2	70·5	74·4	80·0	84·8	89·2	94·4	98·1	106·0	109·2	65
68·3	71·5	75·4	81·1	86·0	90·3	95·6	99·3	107·3	110·5	66
69·3	72·6	76·5	82·2	87·1	91·5	96·8	100·6	108·5	111·7	67
70·3	73·6	77·6	83·3	88·3	92·7	98·0	101·8	109·8	113·0	68
71·3	74·6	78·6	84·4	89·4	93·9	99·2	103·0	111·1	114·3	69
72·4	75·7	79·7	85·5	90·5	95·0	100·4	104·2	112·3	115·6	70
73·4	76·7	80·8	86·6	91·7	96·2	101·6	105·4	113·6	116·9	71
74·4	77·8	81·9	87·7	92·8	97·4	102·8	106·6	114·8	118·1	72
75·4	78·8	82·9	88·8	93·9	98·5	104·0	107·9	116·1	119·4	73
76·4	79·9	84·0	90·0	95·1	99·7	105·2	109·1	117·3	120·7	74
77·5	80·9	85·1	91·1	96·2	100·8	106·4	110·3	118·6	121·9	75
78·5	82·0	86·1	92·2	97·4	102·0	107·6	111·5	119·9	123·2	76
79·5	83·0	87·2	93·3	98·5	103·2	108·8	112·7	121·1	124·5	77
80·5	84·0	88·3	94·4	99·6	104·3	110·0	113·9	122·3	125·7	78
81·5	85·1	89·3	95·5	100·7	105·5	111·1	115·1	123·6	127·0	79
82·6	86·1	90·4	96·6	101·9	106·6	112·3	116·3	124·3	128·3	80
83·6	87·2	91·5	97·7	103·0	107·8	113·5	117·5	126·1	129·5	81
84·6	88·2	92·5	98·8	104·1	108·9	114·7	118·7	127·3	130·8	82
85·6	89·2	93·6	99·9	105·3	110·1	115·9	119·9	128·6	132·0	83
86·6	90·3	94·7	101·0	106·4	111·2	117·1	121·1	129·8	133·3	84
87·7	91·3	95·7	102·1	107·5	112·4	118·2	122·3	131·0	134·5	85
88·7	92·4	96·8	103·2	108·6	113·5	119·4	123·5	132·3	135·8	86
89·7	93·4	97·9	104·3	109·8	114·7	120·6	124·7	133·5	137·0	87
90·7	94·4	98·9	105·4	110·9	115·8	121·8	125·9	134·7	138·3	88
91·7	95·5	100·0	106·5	112·0	117·0	122·9	127·1	136·0	139·5	89
92·8	96·5	101·1	107·6	113·1	118·1	124·1	128·3	137·2	140·8	90
93·8	97·6	102·1	108·7	114·3	119·3	125·3	129·5	138·4	142·0	91
94·8	98·6	103·2	109·8	115·4	120·4	126·5	130·7	139·7	143·3	92
95·8	99·6	104·2	110·9	116·5	121·6	127·6	131·9	140·9	144·5	93
96·8	100·7	105·3	111·9	117·6	122·7	128·8	133·1	142·1	145·8	94
97·9	101·7	106·4	113·0	118·8	123·9	130·0	134·2	143·3	147·0	95
98·9	102·8	107·4	114·1	119·9	125·0	131·1	135·4	144·6	148·2	96
99·9	103·8	108·5	115·2	121·0	126·1	132·3	136·6	145·8	149·5	97
100·9	104·8	109·5	116·3	122·1	127·3	133·5	137·8	147·0	150·7	98
101·9	105·9	110·6	117·4	123·2	128·4	134·6	139·0	148·2	151·9	99
102·9	106·9	111·7	118·5	124·3	129·6	135·8	140·2	149·4	153·2	100

Appendix 3 Generic Failure Rates

The following figures are based on random failures of electronic components and, hence, constant failure rates apply. The failures are assumed to be catastrophic, that is to say, sudden and complete. No allowance for drift failures is made and it is assumed that circuit tolerancing makes due allowance for this factor. It should be borne in mind that poor circuit tolerancing can cause component failure rates ten times higher than quoted.

Generic failure rates are quoted here for the benefit of the reader who wishes to understand the types of values encountered. The fuller detail of MIL 217C runs to over 200 pages. The following, from MIL 217C, are expressed in units of 10^{-6} per hour.

When performing a parts count feasibility prediction, generic failure rates can be used as follows.

$$\lambda = \sum_{1}^{n} N_i(\lambda_g \pi_q)i$$

where N is the number of each part type; λ_g is the generic failure rate at the relevant environment; π_q is the quality multiplying factor.

A description of the symbols for the environmental conditions will be found in section 7.4.

π_Q, Quality factors

Quality level	Description	π_Q
S	Procured in full accordance with MIL-M-38510, Class S requirements	
B	Procured in full accordance with MIL-M-38510, Class B requirements	
B–1	Procured to screening requirements of MIL-STD-883, Method 5004, Class B, and in accordance with the electrical requirements of MIL-M-38510 slash sheet or vendor or contractor electrical parameters. The device must be qualified to requirements of MIL-STD-883, Method 5005, Class B. No waivers are allowed	

B–2   Procured to vendor's equivalent of screening
requirements of MIL-STD-883, Method 5004,
Class B, and in accordance with vendor's
electrical parameters. Vendor waives certain
requirements of MIL-STD-883, Method 5004, Class B

C   Procured in full accordance with MIL-M-38510,
Class C requirements

C–1   Procured to screening requirements of MIL-STD-883,
Method 5004, Class C and in accordance with the
electrical requirements of MIL-M-38510 slash sheet or
vendor or contractor electrical specification. The
device must be qualified to requirements of
MIL-STD-883, Method 5005, Class C. No waivers are
allowed

D   Commercial (or non-mil standard) part, hermetically
sealed, with no screening beyond the manufacturer's
regular quality assurance practices

D–1   Commercial (or non-mil standard) part, packaged or
sealed with organic materials (e.g. epoxy, silicone, or
phenolic)

	Quality level	π_Q
Quality factors for use with tables A and B	S	0.5
	B	1
	B–1	2.5
	B–2	5
	C	8
	C–1	45
	D	75
	D–1	150

For tables A and B a factor of ten (times) should be applied for new devices or
where there have been major changes in production.

Table A

Generic failure rate, λ_G, for bipolar digital devices (TTL and DTL) vs. Environment (f./10^6 h)

Circuit Complexity	G_B and S_F	G_F	A_{IT}	A_{IF}	N_S	G_M	A_{UT}	A_{UF}	N_U	M_L
1–20 gates	.0070	.029	.070	.13	.093	.091	.11	.20	.12	.21
21–50 gates	.020	.062	.12	.21	.17	.16	.20	.33	.23	.34
51–100 gates	.032	.094	.18	.29	.24	.23	.28	.45	.34	.47
101–500 gates	.079	.22	.37	.56	.49	.45	.61	.89	.71	.85
501–1000 gates	.13	.34	.56	.82	.73	.67	.92	1.3	1.1	1.2
1001–2000 gates	.29	.78	1.3	1.8	1.7	1.5	2.1	2.9	2.5	2.7
2001–3000 gates	.81	2.1	3.5	5.1	4.5	4.1	5.8	8.1	6.7	7.3
3001–4000 gates	2.2	5.7	9.6	14.	12.	11.	16.	22.	18.	20.
4001–5000 gates	5.9	16.	26.	38.	33.	30.	43.	60.	49.	54.
ROM*, ⩽ 320 bits	.0083	.022	.036	0.53	.048	.043	.060	.085	.070	.078
ROM* 321–576 bits	.012	.033	.055	.081	.072	.66	.091	.13	.11	.12
ROM* 577–1120 bits	.020	.052	.087	.13	.11	.10	.14	.20	.17	.19
ROM* 1121–2240 bits	.029	.078	.12	.20	.17	.16	.22	.31	.25	.29
ROM* 2241–5000 bits	.045	.12	.20	.30	.27	.24	.33	.48	.39	.45
ROM* 5001–11000bits	.068	.18	.31	.47	.41	.38	.51	.75	.60	.70
ROM 11001–17000 bits	.10	.28	.48	.73	.63	.58	.79	1.1	.92	1.1

*RAM failure rate = 3.5 x ROM failure rates.

Table B

Generic failure rate, λ_G, vs. Environment for bipolar beam lead, ECL, all linear, and all MOS devices (f./10^6 h)

Circuit complexity	G_B and S_F	G_F	A_{IT}	A_{IF}	N_S	G_M	A_{UT}	A_{UF}	N_U	M_L
1–20 gates	.010	.048	.099	.16	.14	.12	.21	.30	.25	.24
21–50 gates	.048	.19	.31	.40	.43	.34	.73	.86	.92	.52
51–100 gates	.076	.31	.48	.59	.68	.54	1.2	1.3	1.5	.78
101–500 gates	.19	.82	1.2	1.4	1.7	1.3	3.1	3.4	3.9	1.7
501–1000 gates	.32	1.4	2.0	2.3	2.8	2.1	5.1	5.5	6.4	2.6
1001–2000 gates	.74	3.1	4.6	5.2	6.4	4.8	12	13.	15.	6.0
2001–3000 gates	2.0	8.4	13.	14.	17.	13.	33.	35.	41.	16.
3001–4000 gates	5.4	23.	35.	39.	47.	36.	90.	96.	111.	44.
4001–5000 gates	15.	62.	94.	105.	128.	97.	241.	258.	299.	121.
ROM*, ≤320 bits	.021	.087	.13	.15	.18	.14	.33	.36	.42	.17
ROM* 321–576 bits	.031	.13	.19	.22	.27	.20	.49	.53	.62	.26
ROM* 577–1120 bits	.048	.20	.31	.35	.42	.32	.78	.84	.98	.41
ROM* 1121–2240 bits	.072	.30	.45	.52	.63	.48	1.2	1.3	1.5	.61
ROM* 2241–5000 bits	.11	.46	.70	.80	.96	.74	1.8	1.9	2.2	.94
ROM* 5001–11000 bits	.17	.70	1.1	1.2	1.5	1.1	2.7	2.9	3.4	1.5
ROM* 11001–17000 bits	.25	1.1	1.6	1.9	2.2	1.7	4.1	4.5	5.2	2.2
Linear, ≤32 transistors	.011	.052	.12	.20	.16	.15	.22	.35	.27	.33
Linear, 33–100 transistors	.023	.11	.24	.41	.35	.31	.48	.73	.60	.66

*RAM failure rate = 3.5 x ROM failure rate.

Table C.1

Generic failure rate, λ_G, (f./10^6 h) for discrete semiconductors vs. Environment (see Table C.2 for quality factor)

Part type	G_B and S_F	G_F	A_{IT}	A_{IF}	N_S	G_M	A_{UT}	A_{UF}	N_U	M_L
Transistors										
Si NPN	.017	.11	.28	.59	.26	.59	.60	1.2	.84	.96
	.025	.17	.46	.96	.41	.96	.96	1.9	1.4	1.5
Si PNP										
Ge PNP	.025	.25	.75	1.6	.84	1.6	.78*	1.6*	2.1*	2.5
Ge NPN	.072	.66	2.0	4.3	2.2	4.3	3.3*	6.6*	5.4*	6.6
FET	.046	.31	.78	1.6	.70	1.6	1.7	3.4	2.3	2.6
Unijunction	.15	1.0	2.7	5.6	2.4	5.6	6.3	13.	9.0	9.0
Diodes										
Si, gen. purpose	.0051	.036	.098	.20	.090	.20	.24	.48	.33	.33
Ge. gen purpose	.0066	.078	.25	.51	.30	.51	.44	.87*	.75*	.81
Zener & Avalanche	.016	.096	.24	.51	.22	.51	.54	1.1	.72	.84
Thyristor	.023	.16	.43	.90	.40	.90	1.0	2.0	1.4	1.4
Si Microwave Det.	.19	2.2	6.0	12.	3.9	12.	7.5	25.	17.	46.
Ge Microwave Det.	.41	5.6*	18.*	35.*	**	35.*	**	**	**	**
Si Microwave Mix.	.25	3.0	8.0	16.	5.1	16.	17.	34.	23.	64.
Ge Microwave Mix.	.72	10.*	31.*	61.*	**	61.*	**	**	**	**
Varactor, Step Recovery, Tunnel	.24	1.5	3.9	8.1	3.5	8.1	8.6	17.	12.	13.
LED	.034	.14	.25	.49	.45	.35	.91	1.8	1.4	.88
Single Isolator	.051	.21	.38	.74	.68	.53	1.4	2.7	2.1	1.30

* This value is valid only for electrical stress, $S \le 0.3$.

** Do not use in these environments since temperature normally encountered combined with normal power dissipation are above the device ratings.

Table C.2

π_Q, quality factors for Table C.1

Part type	JANTXV	JANTX	JAN	NON-MIL HERMETIC	PLASTIC
Microwave Diodes	0.3	0.6	1.0	2.0	—
All Other Types	0.1	0.2	1.0	5.0	10.

Table D

Generic failure rate, λ_G, (f./10^6 h) for resistors

Resistors, fixed								Use environment				
Construction	Style	MIL-R-Spec.	G_B and S_F	G_F	A_{IT}	A_{IF}	N_S	G_M	A_{UT}	A_{UF}	N_U	M_L
Composition	RCR	39008	.00051	.0032	.0037	.0075	.0046	.0066	.014	.027	.021	.02
Composition	RC	11	.0025	.016	.018	.038	.023	.033	.069	.13	.10	.099
Film	RLR	39017	.0012	.0031	.0043	.0088	.0033	.0062	.016	.032	.021	.028
Film	RL	22684	.0061	.015	.022	.044	.017	.031	.079	.016	.10	.14
Film	RN	55182	.0014	.0033	.0049	.01	.0037	.007	.018	.036	.024	.032
Film	RN	10509	.0073	.017	.025	.05	.019	.035	.09	.18	.12	.16
Film, power	RD	11804	.012	.026	.055	.11	.026	.078	.15	.29	.19	.46
Film, network	RZ	83401	.026	.072	.17	.34	.14	.24	.80	1.6	1.2	1.1
Wirewound, accurate	RBR	39005	.0085	.019	.058	.12	.020	.078	.22	.44	.17	.39
	RB	93	.043	.094	.29	.58	.10	.39	1.1	2.2	.85	1.9
Wirewound, power	RWR	39007	.014	.044	.073	.15	.037	.091	.19	.37	.25	.54
	RW	26	.072	.22	.36	.73	.19	.45	.94	1.9	1.3	2.7
Wirewound, Ch. Mount	RER	39009	.0079	.021	.045	.090	.024	.056	.11	.22	.16	.34
	RE	18546	.040	.010	.22	.45	.12	.28	.56	1.1	.79	1.7

Resistors, Variable

Wirewound, trimmer	RTR	39015	.014	.034	.078	.16	.077	.15	.19	.37	.24	1.1
	RT	27208	.072	.17	.39	.79	.39	.74	.93	1.9	1.2	5.5
W. W., Prec.	RR	12934	.84	2.1	5.5	11.	4.6	11.	14.	29.	18.	132.
W. W., semi-prec.	RA	19	.31	.84	2.4	4.8	2.1	9.5	*	*	*	*
	RK	39002	"	"	"	"	"	"	*	*	*	*
W. W., power	RP	22	.31	.78	2.0	3.9	1.7	7.8	*	*	*	*
Non-W. W. trimmer	RJR	39035	.02	.067	.12	.23	.95	.23	.27	.54	.34	1.8
	RJ	22097	.10	.33	.58	1.2	4.8	1.2	1.4	2.7	1.7	9.2
Composition	RV	94	.12	.49	1.1	2.1	.88	4.1	6.6	13.	6.8	18.
Non-WW. prec.	RQ	39023	.086	.35	.65	1.3	.58	1.3	2.8	5.7	2.6	10.
Film	RVC	23285	.096	.34	.6	1.2	.51	1.2	2.2	4.4	2.0	9.6

* Not normally used in these environments.

Table E

Generic failure rate, λ_G, (f./10^6 h) for capacitors

Capacitors, fixed

Dielectric	Style	MIL-C-Spec.	G_B and S_F	G_F	A_{IT}	A_{IF}	N_S	G_M	A_{UT}	A_{UF}	M_U	M_L
							Use environment					
Paper	CP	25	.011	.022	.057	.11	.047	.057	.16	.31	.15	.29
Paper	CA	12889	.012	.031	.087	.17	.083	.087	.47	.90	.53	.44
Paper/Plastic	CZR	11693	.0047	.0098	.025	.05	.021	.025	.072	.14	.064	.13
Paper/Plastic	CPV	14157	.0021	.0042	.0088	.018	.0088	.0088	.025	.05	.023	.044
Paper/Plastic	CQR	19978	"	"	"	"	"	"	"	"	"	"
Paper/Plastic	CHR	39022	.0028	.006	.012	.024	.012	.012	.035	.069	.032	.06
Paper/Plastic	CH	18312	.02	.042	.084	.17	.087	.084	.24	.49	.22	.42
Plastic	CFR	55514	.0041	.0086	.022	.043	.018	.03	.067	.13	.075	.13
Plastic	CRH	83421	.0023	.0048	.0096	.019	.010	.0096	.028	.055	.025	.048
Mica	CMR	39001	.0005	.0022	.0059	.012	.0043	.0084	.044	.088	.042	.042
Mica	CM	5	.003	.013	.035	.071	.026	.050	.27	.53	.25	.25
Mica	CB	10950	.09	.19	.42	.85	.3	.60	1.9	3.8	1.4	3.0

Glass	CYR	23269	.0003	.0014	.0037	.0075	.0066	.0053	.027	.054	.028	.026
Glass	CY	11272	.001	.0043	.011	.022	.020	.016	.082	.16	.084	.079
Ceramic	CKR	39014	.0036	.0076	.033	.066	.0098	.016	.068	.14	.032	.12
Ceramic	CK	11015	.011	.023	.099	.20	.029	.047	.20	.41	.096	.35
Ceramic	CCR	20	.0008	.0032	.008	.016	.0058	.011	.058	.12	.070	.057
TA, SOL.	CSR	39003	.012	.026	.078	.16	.035	.052	.15	.29	.14	.26
TA, Non-Sol	CLR	39006	.0061	.014	.082	.16	.049	.069	.14	.28	.15	.23
TA, Non-Sol	CL	3965	.018	.043	.24	.49	.15	.21	.42	.83	.46	.69
Al oxide	CU	39018	.074	.23	1.2	2.3	.96	1.6	4.8	9.7	5.3	5.5
Al dry	CE	62	.090	.36	1.9	3.7	1.7	2.6	10.	21.	12.	8.7

Capacitors, variable

Ceramic	CV	81	.32	1.6	2.5	4.8	4.2	3.5	24.	48.	19.	31.
Piston	PC	14409	.099	.54	1.2	2.3	1.7	1.5	9.2	18.	22.	7.5
Air, trimmer	CT	92	.4	3.0	4.8	9.4	8.	6.8	49.	98.	37.	60.
Vacuum	CG	23183	1.2	6.2	15.	29.	15.	21.	140.	270.	94.	*

* Not normally used in this environment.
For Tables D and E use $\pi_q = 1$ for Military components and $\pi_q = 3$ for commercial.

Table F

Generic failure rate, λ_G, (f./10^6 h) for inductive, electromechanical and miscellaneous parts

Part type	G_B and S_F	G_F	A_{IT}	A_{IF}	N_S	G_M	A_{UT}	A_{UF}	N_U	M_L
					Use environment					
Inductive										
Low power pulse transformer	.003	.0048	.041	.082	.017	.047	.069	.14	.065	.12
Audio transformer	.006	.0096	.082	.16	.034	.094	.14	.28	.13	.24
High power pulse and power transformer filter	.019	.053	.31	.60	.13	.35	.46	.92	.98	.86
R. F. transformer	.024	.038	.33	.64	.14	.38	.56	1.1	.52	.96
R. F. coils, fixed	.0016	.004	.021	.042	.0096	.048	.039	.078	.038	.12
F. R. coils, variable	.0032	.008	.042	.084	.019	.019	.078	.16	.077	.24
Motors	*	15.	19.	19.	24.	19.	41.	41.	49.	*
Relays										
General purpose	.13	.30	.65	1.3	.89	.81	2.8	5.6	2.9	16.
Contacter, high current	.44	1.0	2.2	4.5	3.0	2.8	9.6	19.	10.	56.
Latching	.10	.24	.52	1.0	.71	.65	2.2	4.5	2.3	13.
Reed	.11	.26	.55	1.1	.75	.69	2.4	4.8	2.5	14.
Thermal bi-metal	.29	.69	1.5	3.0	2.0	1.9	6.4	13.	6.7	37.
Meter movement	.90	2.1	4.6	9.2	6.3	5.8	20.	40.	21.	*

Switches

Toggle and push button									
.035	.011	.18	.35	.15	.61	1.8	3.5	.84	24.
Sensitive									
.15	.44	.74	1.5	.59	2.5	7.4	15.	3.4	100.
Rotary									
.22	.67	1.1	2.2	.89	3.8	11	22.	5.1	150.

Connectors (per pair)

Circular, rack and panel									
.0062	.029	.12	.24	.053	.12	.17	.34	.23	.18
Printed wiring board									
.0031	.028	.060	.12	.036	.060	.090	.18	.11	.090
Coaxial									
.0084	.032	.13	.26	.060	.10	.18	.36	.24	.20

P. C. wiring boards

Two-sided									
.0012	.0024	.005	.01	.0048	.0048	.012	.024	.012	.024
Multi-layer									
.15	.30	.63	1.3	.60	.60	1.5	3.0	1.5	3.0

* Not normally used in these environments.

π_Q **Factor for use with Table F**

Part type	Quality level	
	MIL-SPEC	Non-MIL
Inductive	1	3
Motors	1	1
Relays	1	3
Switches, toggle and sensitive	1	20
Switches, rotary	1	50
Connectors	1	3
P.W. boards	1	—
Others	—	1

Appendix 4 Terotechnology

Terotechnology is another word for the overall management of resources and costs over the life cycle of a machine, plant, or process. The aim is to achieve the right balance of activities and expenditure so that the overall total cost of installation, maintenance, modification, replacements, etc., is brought to a minimum.

The concept has been promoted during the 1970s in the UK by the Department of Industry. The term implies both mystique and complexity but is essentially a practical and integrated approach to resource management. One desirable effect is to bring maintenance management closer to the design process thereby bridging a long standing gap. It is to be hoped that the practical appeal of such an approach will find more rapid acceptance than did Reliability and Maintainability despite the clumsiness of the term. It is fundamental to a Terotechnology approach that comprehensive feedback data from all stages between design and eventual replacement is channelled to a central point for analysis. Historical information can then be used as an input to Life Cycle Costing. This involves a study of the total Cost of Ownership, of a plant or equipment, throughout its operating life. Life cycle costs include the cost of consumeable items, preventive and corrective maintenance, and spares. The accuracy of such a study is dependent on the accuracy of MTBF and MTTR predictions.

The subject is still new although it involves merely a modern approach to established functions. The principles of cost analysis and optimisation, as discussed in chapter 2, if applied in the widest possible context would provide the main input, and ultimately the justification, for a Terotechnology study.

Appendix 5 Bibliography

RELIABILITY

Bazovsky, I., *Reliability Theory and Practice*, 2nd edn, Prentice Hall, New York, 1971

Caplan, *A Practical Approach to Reliability*, Business Books, London, 1972

Dummer, G. W., *Elementary Guide to Reliability*, 2nd edn, Oxford University Press, 1974

Kivenson, G., *Durability and Reliability in Engineering Design*, Pitman, London, 1972

Myers, Wong and Gordy, *Reliability Engineering for Electronic Systems*, Wiley, New York, 1964

Smith, D. J., *Reliability Engineering*, Pitman, London, 1972

Staff of ARINC, *Reliability Engineering*, Prentice Hall, New York, 1964

MAINTAINABILITY

Blanchard and Lowery, *Maintainability Principles and Practice*, McGraw-Hill, New York, 1969

Goldman and Slattery, *Maintainability – A Major Element of System Effectiveness*, Wiley, New York, 1964

Smith and Babb, *Maintainability Engineering*, Pitman, London, 1973

STATISTICS

Moroney, M. J., *Facts from Figures*, Pelican, London, 1951

Snedecor and Cochran, *Statistical Methods*, Iowa State University Press, 1967

Smith, D. J., *Statistics Workshop*, Technis, 1974

TEROTECHNOLOGY

Husband, *Maintenance Management and Terotechnology*, Saxon House, 1976

Various authors, *Terotechnology Handbook*, Dept. of Industry, 1978

US MILITARY HANDBOOKS AND STANDARDS

MIL HDBK 217C : Reliability Prediction of Electronic Equipment (1979)
                (N.B. Contains failure rate data. Update of 217B (1974).)
MIL STD 470   : Maintainability Programme Requirements (1966)
MIL STD 471A   : Maintainability Verification/Demonstration/Evaluation
                (1973)
MIL HDBK 472   : Maintainability Prediction (1961)
MIL STD 721B   : Definitions of Effectiveness Terms for Reliability,
                Maintainability, Human Factors and Safety
MIL STD 781B   : Reliability Test: Exponential Distribution (1967)
MIL STD 785A   : Reliability Programme for Systems and Equipment
                Development and Production (1969)

OTHER STANDARDS

British Standard 2011 : Basic Environmental Testing Procedures
                        (N.B. This is in 45 Sections)
British Standard 4200 : Guide on the Reliability of Electronic Equipment and
                        Parts used therein
British Standard 4778 : Glossary of Terms used in Quality Assurance
                        (N.B. Includes reliability and maintainability)
British Standard 5760 : Reliability of Systems, Equipment and Components
IEC Publication   271 : Preliminary List of Basic Terms and Definitions for the
                        Reliability of Electronic Equipment and the Components
                        Components (or parts) used therein
Defence Standard 00-5, Parts 1–3, Issue 3:
                        Design and Construction Criteria for Reliability and
                        Maintainability of Land Service Materiel

Appendix 6 Answers to Exercises

Chapter 12

1. Accumulated time $T = 50 \times 100 = 5000$ h. Since the test was time truncated $n = 2(k + 1)$ Therefore,

   (a) $n = 6, T = 5000,$ $\alpha = 0.4.$ From appendix 2, $\chi^2 = 6.21$

   $$\text{MTBF}_{60\%} = \frac{2T}{\chi^2} = \frac{10\,000}{6.21} = 1610 \text{ h}$$

   (b) $n = 2, T = 5,000,$ $\alpha = 0.4.$ From appendix 2, $\chi^2 = 1.83$

   $$\text{MTBF}_{60\%} = \frac{2T}{\chi^2} = \frac{10\,000}{1.83} = 5464 \text{ h}$$

2. If $k = 0$ then $n = 2$ and since confidence level $= 90\%$ $\alpha = 0.1$

   Therefore $\chi^2 = 4.61$

   $$\text{MTBF}_{90\%} = 5000 = \frac{2T}{\chi^2} = \frac{2T}{4.61}$$

   Therefore $T = \dfrac{5000 \times 4.61}{2} = 11\,525 \text{ h}$

   Since there are 50 devices the duration of the test is $\dfrac{11525}{50} = 231$ h.

3. From figure 12.7. If $c = 0$ and $P_{0-c} = 0.85$ $(\alpha = 0.15)$ then $m = 0.17$

   Therefore $T = m\theta = 0.17 \times 1000 = 170$ h

   If MTBF is 500 h then $m = T/\theta = 170/500 = 0.34$ which shows $\beta = 70$ per cent

   If $c = 5$ then $m = 3.6$ at $P_{0-c} = 0.85$

   Therefore $T = m\theta = 3.6 \times 1000 = 3600$ h

   If MTBF is 500 h then $m = T/\theta = 3600/500 = 7.2$ which shows $\beta = 28$ per cent

   N.B. Do not confuse α meaning $(1 - \text{confidence level})$ with α as producer's risk.

Chapter 13

1. From the example $R(t) = \exp\left[\left(\dfrac{-t}{1110}\right)^{1.5}\right]$

   If $R(t) = 0.95$ Then $\left(\dfrac{t}{1110}\right)^{1.5} = 0.051$

   Therefore $1.5 \log t/1110 = \log 0.051$

   Therefore $\log t/1110 = -1.984$

   Therefore $t/1110 = 0.138$

   Therefore $t = 153$ h

2. Using the table of median ranks, sample size 10, as given in chapter 13, plot the data and verify that a straight line is obtained.

   Note that $\beta = 2$ and that $\eta = 13\,000$ h

   Therefore

   $$R(t) = \exp\left[\left(\dfrac{-t}{13\,000}\right)^{2}\right]$$

   and $MTBF = 0.886 \times 13\,000 = 11\,500$ h

Chapter 15

1. $R(t) = e^{-\lambda t}\,[2e^{-\lambda t} - e^{-2\lambda t}]$

   $= 2e^{-2\lambda t} - e^{-3\lambda t}$

   $MTBF = \displaystyle\int_{0}^{\infty} R(t)\,dt = \dfrac{1}{\lambda} - \dfrac{1}{3\lambda} = \dfrac{2}{3\lambda}$

   N.B. Not a constant failure rate system despite λ being constant.

2. This is a conditional redundancy problem. Consider the reliability of the system if (a) B does fail and (b) B does not fail. The following two block diagrams describe the equivalent systems for these two possibilities.

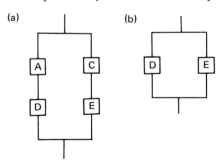

Using Bayes theorem the reliability is given as:

Reliability of diagram (a) $\times$ probability that B fails (i.e. $1 - R_b$)

PLUS

Reliability of diagram (b) $\times$ probability that B does not fail (i.e. R_b)

Therefore System Reliability

$$= [RaRd + RcRe - RaRdRcRe] \, (1 - Rb) + [Rd + Re - RdRe] \, Rb$$

3. Since $R(t)$ of the system without repair is $2e^{-2\lambda t} - e^{-3\lambda t}$

$$
\begin{aligned}
\text{MTBF} &= \frac{\displaystyle\int_0^T R(t)\,dt}{1 - R(T)} \\[2em]
&= \frac{\displaystyle\int_0^T (2e^{-2\lambda t} - e^{-3\lambda t})\,dt}{1 - 2e^{-2\lambda T} + e^{-3\lambda T}} \\[2em]
&= \frac{\left[\dfrac{-e^{-2\lambda t}}{\lambda} + \dfrac{e^{-3\lambda t}}{3\lambda}\right]_0^{T}}{1 - 2e^{-2\lambda T} + e^{-3\lambda T}} \\[2em]
&= \frac{e^{-3\lambda T} - 3e^{-2\lambda T} + 2}{3\lambda(1 - 2e^{-2\lambda T} + e^{-3\lambda T})}
\end{aligned}
$$

Index